GREAT FIRES
OF
AMERICA

GREAT FIRES
OF AMERICA

By the Editors of Country Beautiful

Publisher and Editorial Director: Michael P. Dineen
Executive Editor: Robert L. Polley
Managing Editor: John M. Nuhn
Text by Nancy Backes

Country Beautiful Corporation

Waukesha, Wisconsin

This book is dedicated to America's heroic firemen who relentlessly battle fires that destroy lives and property, sometimes losing their own in the effort.

COUNTRY BEAUTIFUL: *Publisher and Editorial Director:* Michael P. Dineen; *Executive Editor:* Robert L. Polley; *Senior Editors:* Kenneth L. Schmitz, James H. Robb, Stewart L. Udall; *Art Director:* Buford Nixon; *Managing Editor:* John M. Nuhn; *Associate Editors:* D'Arlyn M. Marks, Kay Kundinger; *Assistant Editor:* Nancy Backes; *Production Manager:* Donna Griesemer; *Administration:* Brett E. Gries, Bruce L. Schneider.

Country Beautiful Corporation is a wholly-owned subsidiary of the Flick-Reedy Corporation: *President:* Frank Flick; *Vice President and General Manager:* Michael P. Dineen; *Treasurer and Secretary:* August Caamano.

ACKNOWLEDGEMENTS

The Editors wish to thank the following whose invaluable assistance in the production of this book is greatly appreciated: Paul Clifford of Wide World Photos; fire fighter and photographer James Haight, Milwaukee Fire Department; Mary Frances Rhyner and Julia Cracraft of the Chicago Historical Society; George Getz and Janice Lippmann of the Hall of Flame, Scottsdale, Arizona; George Anne Daly of the Insurance Company of North America Archives Department; George Talbot and Melodie Knisely of the State Historical Society of Wisconsin; Chief William Stamm, Deputy Chief Richard Seelen and Supervisor-Coordinator of Fire Technology George McDonald, all of the Milwaukee Fire Department; fire fighter Bill Reynolds of the Elmhurst, Illinois, Fire Department; John Ottoson of the National Fire Protection Association; Don Davidson of the *Newark* (New Jersey) *Star-Ledger*, and all libraries, newspapers and private collectors throughout the country who spent much time and effort locating materials.

Pages 2-3: Fire produces curious lighting effects in the night air as fire fighters and their modern equipment advance on the flames. Robert Bartosz. *Frontispiece:* Flames appear in nearly every window as Cyrus McCormick's Reaper Works and the Chicago Steam Sugar Refinery succumb to the ruinous Great Chicago Fire of 1871. Chicago Historical Society. *Right:* Though fire and smoke often seriously threaten fire fighters' lives, their first obligation is to enter the burning building and search for trapped life. Joseph Mancinelli, Detroit Fire Department.

Contents

I	Something Burns Every Day	10
II	Early America in Ashes	26
III	From Political Revolution to Industrial Revolution	32
IV	Conflagration in Chicago	48
V	The Heavens Rained Fire in the North Woods	60
VI	Colorado Mining: Fortune and Fire	76
VII	Earthquake and Flames	82
VIII	Fire On the Waterfront	92
IX	SOS at Sea	100
X	The Triangle Shirtwaist Tragedy	106
XI	Hotel Holocausts	114
XII	School Bells and Fire Bells	124
XIII	Theaters: The Show Did Not Go On	134
XIV	Government-Sponsored Firetraps	142
XV	Devastation at the Cocoanut Grove	146
XVI	Fire Under the Big Top	154
XVII	Gushers of Flame	164
XVIII	Hospitals: Healing and Dying	170
XIX	Burn, Baby, Burn	176
XX	Firemen in a Modern America	194
Appendix — Fire Museums in the U.S.		206

The quick clomping of spirited, dappled gray horses, thick smoke pouring from the steamer's shiny stack, and the "always ready" firemen spelled excitement for the population at the turn of the century. Crowds gathering to watch promptly got out of the way. Pabst Brewing Company.

Something Burns Every Day

It is a dark, stormy night in the year 1,000,000 B.C. A curious form of life, half ape and half human, stands incredulously watching the tempest before him. The environment only an hour before was peaceful and serene. Now it is a churning fury. Suddenly there is a brilliant, jagged white flash in the sky eerily illuminating the area, followed by a loud, trembling clap. The tree directly beneath the jagged flash has taken up a fearful, bright red and orange glow. The flickers are hot tongues of light that crackle as they devour the familiar trees, bushes and grass. The more it consumes, the larger and more awesome it becomes. It lights the sky and becomes the toy of the wind. The once tranquil forest is illuminated with its raging, contagious glow. Humanity is introduced to Fire.

"Man is the animal that has made friends with the fire" observed Henry van Dyke. Yet it was quite some time before humanity learned that this destroyer, this ravager of the land, could be its friend. But eventually fire was conquered. We know today that it was the making of tools and the mastering of fire that distinguished early humans from other animals. It was fire that lit humanity's way out of the tropics into other lands. It was fire that enabled early humans to make better tools. And it was fire that introduced gourmet treats other than bark and berries, increasing the amount of digestible foods. The hearth became the center of the home and even today it is popular to think of happy homes being quietly settled around a fireplace.

The art of firemaking is 25,000 to 50,000 years old. Peking Man was perhaps the first regular user of fire. Neanderthal Man learned to manufacture the phenomenon. Fire was at the center of human culture. Humans learned that if they set fire to the brush, they would have better gamelands. The Great Plains of America are one of the results of these fires. They learned that ashes made good fertilizer for their plants. And in an attack by wild animals or other humans, fire was an excellent weapon. Its mysterious glow was almost instinctively feared. Fire was a successful tool in warfare for no one could withstand its destructiveness. All knew that fire, though an incredible, benevolent tool, would, if neglected or misused, become a burning enemy.

Burnings have been an effective weapon in history. The Romans were not ignorant of this, and staged many attacks with the use of fire. Ironically, Rome itself was destined to burn. A devastating fire in 12 B.C. was apparently enough to cause Emperor Augus-

Eskimos (above) made fire with iron and flint to heat their food, while the Incas of South America (overleaf) used fire in ceremonial worship. A warm hearth (pages 14-15) comforted colonial travelers in America. All paintings from the Universal Match collection.

tus to take steps for future fire protection. Seven thousand freemen were organized into seven battalions. Augustus, wise in the areas of finance, knew that such an extensive service required funds, so a twenty-five percent tax on slaves was imposed. The members of the department were also trained in fire prevention as well as protection and regularly made "careful inspections of the kitchens, of the heating apparatus, and of the water supplies in the houses,

and every fire was the subject of judicial examination." These people were forerunners to the modern building inspectors.

A favorite practice in antiquity and the middle ages was the use of fireships, whereby a ship was set on fire and floated toward enemy ships to envelop them with this luminous and impartial killer. The tactic was used in the United States as late as April 1862, during the Civil War, when the Confederate flagship, *Hart-*

(continued on page 16)

ford was set afire by a flaming skeleton hulk. The fires of World War II were more awesome than when the first human beings saw lightning strike a tree in a storm. This time human beings were the fire gods attacking appropriately from the sky, and the storm they created was not composed of wind and rain, but of fire. In a fire storm, created by an atomic bomb, the intense heat crumbles buildings beneath it. The air itself appears to be of fire. There is no escape. At Hiroshima, Japan, in 1945, after the atomic bomb exploded, a fire storm indiscriminately consumed seventy to eighty thousand people.

Anyone who has sat before a fire and listened to its gentle crackling and contemplated near its friendly warmth knows something of the ancients' respect for the fire. The little fire that warmed the cave was transformed to a temple of fire worship that commanded the respect and idolation of all people. Fire was so revered, and so necessary, that great care was taken that its magic embers never die. All tribal fires were born from the sacred fire. Often it was the duty of women to see that the eternal fire never went out. Fire rites, some with sacrifices, were held each year by some tribes. The ancient religion of Iran, Zoroastrianism, worshipped fire as a mysterious and awesome phenomenon. Fire was regarded by them to be the most sacred power presented from heaven and kindled by the Deity.

Where did this mystery come from? The ancients could never accept that it was an accident given to humanity by lightning. They reasoned that only the gods would have the power to give such magic to mere mortals. The Greeks believed that fire was stolen from the gods, that Prometheus took fire from the sun's chariot to benefit man. They extolled it as one of the four elements — fire, water, earth and air — that Empedocles said were used to create the world.

The gods' secret of firemaking was kept until someone discovered that by rubbing two sticks together a spark could result. Human ingenuity, always seeking a quicker way, then came up with a device called a fire drill, by which a pointed stick of hardwood was driven into a hole in softer wood to produce a fire.

Since a fire drill could not conveniently be stuffed in a workingman's back pocket, chemical matches were invented in 1805. In 1827 the English chemist, John Walker, invented the friction match, essentially the same match we use today. A quick snap and the oven is lit. A quick snap and the enquiring scientist lights a Bunsen burner. In fact, fire became the symbol of learning as depicted in the lamp of knowledge.

But there was no rejoicing in the streets at its discovery. Americans had for many years helplessly watched their little wooden towns burn and be rebuilt and burn again due to the careless use of matches that were so vulnerable to ignition. The match was dubbed "lucifer match" as if the fire it contained came from the depths of hell, and not the friendly skies of the gods. One professor warned that Americans left themselves wide open for trouble when they traded in their flint and steel for the match. Why, he lamented, a mere footfall or the penetration of the sun's rays, or even a mouse's tooth could hatch a major catastrophe. The lament ended in 1855 with the invention of the safety match by J. E. Lindstrom.

The word "fire" is derived from the Greek word *pyr* which is allied to the Sanskrit *pu*, which means to purify. The Parsees of Bombay today regard fire as a symbol of purity and goodness. Fire has long been recognized as a purifying element, as evidenced by the many pre-modern cultures that maintain an ordeal by fire. A person may be asked, when accused of a crime, to undergo trial by fire. The person may be asked to walk over hot embers, through a blazing log or on hot stones. The embers may be poured over the person's head in a rain of fire. Just as justice is blind, according to the theory, so does fire blindly illuminate and only the guilty are burned. And the majestic fire chooses to burn the people who indulge in this practice much less frequently than would be expected, a fact which begs explanation.

If fire is the purifying element, then America must rank among the most purified countries of the world. Early America, with its narrow streets and wood construction, was destined to burn, and burn again. All America seemed to be built of wood, and the danger of fire was early recognized. Governor Peter Stuyvesant of New Amsterdam (today New York) ordered fire wardens to patrol the streets of Manhattan at night. The wardens were distinguished by their long capes and noisemakers, and the watch became known as the Rattle Watch.

A New Amsterdam law of 1648 ordered that no more wooden chimneys be built, and Boston passed a law in 1654 that required each householder to have a ladder long enough to reach the ridge of his roof and a pole "about 12 feet long, with a good large swob at the end of it." Laws were enforced to keep buckets, bags and fire apparatus in readiness for use. Fire fighting was everyone's concern. As one observer

Livid white flames illuminate the disintegrating skeleton of a frame building and show that fire can be an enemy as well as a necessary tool of civilization. The red clouds of flame that brighten the night sky consume valuable property, and do billions of dollars worth of damage yearly. James Haight.

said: "... the able-bodied man or boy who failed to obey the first summons of the alarm bell, and to work to the best of his ability till the fire was over, would have utterly lost caste."

When America was being settled, fire was intentionally used against nature: the thick forests that covered the land and allowed too little of it to be used for houses and farms. There was a solution — burn it and clear the land. There was so much forest and so few settlers; surely, a few burnings could not amount to much. So immigrants burned, cleared and

built their homes, mostly of wood and close together. They built every house a different height, a different style, creating different problems. Each house was as unique as the idea of the individual, the idea which the country proclaimed to the world.

Fires, however, were not unique. They occurred with a predictable sameness, over and over again. As the threat of destruction grew, so did the concern for danger. Said one citizen, "Some Americans now attribute the frequency of fires in their country partly to the haste, and consequent imperfection of

Opposite: *Winter blazes have been known to be unusually tough ones. Here, fire fighters' streams crystallized to form an intricate tapestry on the building's side. James Haight.* Above: *Quick response to an alarm is essential. In this 1890 photograph a fireman on the truck is still struggling with his boots as children, traditional engine-chasers, race alongside. New Haven Colony Historical Society.*

house building, by which chimneys set fire to the whole dwelling, and yet more to the use of wood for fuel, and the consequent carrying about of wood-ashes."

A survey comparing fires in New York to fires in London in the year 1855 stated that in London there was one fire to every 323 buildings while in New York there was one to every 146 buildings. As if the high combustibility of wood was not enough, America was built of wood in *haste*. There was a noticeable lack of building codes, and Americans in their quickly built dwellings were no different than their ancestors in their fascination with fire. A New Yorker said in 1856, "Political victories achieved and political victories hoped, events abroad and events at home, matters universal, national or local in their interest, all demand the magnificent play of fire."

Fire was colorful. Dangerous? Well, yes, that too.

Yet if America was hasty, America was also ingenious. It could constantly rebuild itself. One British visitor commented in 1878 that "no sooner were the flames extinguished in the burnt district than the occupiers of the premises put up notices on their lots stating their present and future plans.... On the morning after a fire in New York, we were amused in observing that workmen were already engaged in preparations for a new building." All that was required was a little money to build again — and sometimes watch it burn again. Fire was, after all, a price of civilization. Said one, "Do we mean that there shall ever be an end to deaths by fire?"

Early fire departments were strictly on a volunteer basis. Equipment was limited and highly prized. Benjamin Franklin organized the first volunteer fire company in Philadelphia in 1736. Each member was required to furnish, at his own expense, six buckets and two strong linen bags. The company was called The Union Company. In 1730 New York City imported two fire engines from Holland. George Washington was a member of the Volunteer Fire Company of Alexandria, Virginia, which in 1775 purchased a fire engine. In 1852 the first telegraphic fire alarm system was set up in Boston. This is essentially the same system we have today. Cincinnati began the nation's first salaried fire department in 1853, and in 1865 New York abolished its volunteer companies.

At first all equipment was hand operated. Men were not only required to operate the hand pump, a job that was at best clumsy and arduous, but they also were the means of locomotion that moved the equipment to the fire. One engine used in New York in the 1850's weighed nearly 4,500 pounds and required sixteen men to operate the pump. Around the middle of the nineteenth century, hand pumps

began to be replaced by steam pumps. Someone got the idea that horses might be a more expedient means of transportation.

But still there were many obstacles to defy a blaze effectively. After an alarm was turned in, the firehouse door was opened. The horses obediently took their places and waited for the harness to be dropped on their backs from above. Fire fighters who had been asleep upstairs, fully clothed except for boots, were now awake and ready to board their apparatus. Foreign visitors remarked about "those Yankees" that they could be off to a fire within thirty seconds of receiving an alarm. They were fast — although maybe not that fast — but then, they had to be because there was always something on fire. There were, however, serious problems before the fire fighters could reach the fire — such as how to pass through the narrow streets. And how to pass the people, since the fire department did not have the right of way. And, most important — to the fire fighter, anyway — how to be the first company to put

Opposite: *A pair of firemen crouch to the ground as they direct their steady, pressurized stream in an effort to douse an unfriendly blaze. Robert Bartosz.* Above: *Huge, billowy purple clouds of smoke make the firemen in the snorkel seem suspended in mid air, miles from reality. Robert Preudhomme.*

water on that blaze. In those days, rivalry to draw the first water was strong. It was acceptable to trip firemen of other companies, mangle their hose or otherwise prevent them from the glory of proclaiming, "First water!" Many a disappointed homeowner stood before his burning home watching the contest.

Such was the glory of the mid-nineteenth-century fire fighters. And such was the disdain of the public, especially those citizens who had burning homes. Firemen were rugged people glorified in song, poem and lithograph. But they were somehow, according to public opinion, a little removed from civilization — and they were underpaid. "This supplementary payment," remarked one citizen, "invites precisely the sort of persons who can best afford to be firemen, namely, young men with no definite trade or occupation, with a strong love for frolic and adventure, without family ties, in fine, such as live mainly for and in the fresh experience of the passing hour — Bohemians, if we may use the term where neither art nor literature forms a part of its meaning."

In a country founded on seriousness and industriousness, fire fighters were regarded as colorful rogues, and un-American rogues at that, until they were needed, or until no one but the person who made a career of the science of putting out fires could put the fire out. Or until the escape ladder failed, and fire fighter Herman Stauss, unable to listen complacently to cries for help, dared to find a way to save sixteen people stranded on the fifth floor of a burning hotel. There is a tendency to think of fire fighters today, if they are thought of at all, as people who spend a lot of time playing cards and telling stories of the Big One, and just waiting for an alarm. Until we need them.

The threat of fires in business districts, even after departments were modernized, was an enormous one. The principle behind Shakespeare's line, "A little fire is quickly trodden out," was well-understood. Employees at the outbreak of a fire were often found covering the roofs and exposed walls of the building with wet blankets and shawls, while the fire department was fighting for the right of way on some street or competing for first water outside. Sprinkler systems and private fire departments serve the same function today.

Fire occurred daily and took its toll. Why? Who could it be blamed on? Insurance was one target. There was widespread belief in the last century that people were inclined to set fire to their own property. The temptation was certainly there. Less-than-careful people were allowed to obtain insurance and policies were not voided on proof of negligence. In fact, it was to the insurer's disadvantage to discover negligence, since a competitor would inevitably snatch the policy away. Insurance agents too often looked the other way from firetraps. Such was not the case in Europe where English and French common laws provided that a property owner was liable for his neighbor's property should a fire spread.

Fire insurance was a result of the Great Fire of London in 1666, which left four-fifths of the city in ruins. Before insurance, people were largely dependent on charity after a fire. Each insurance company had its fire mark which was posted on the insured property. When an insurance brigade arrived at a fire, they first checked to see if the property bore the fire mark of their company. If it did, the brigade eagerly went to work to save the property. If another mark was found, the brigade quietly stepped aside or slipped back to its quarters.

The first insurance company in America was the Philadelphia Contributionship founded by Benjamin Franklin. The Mutual Assurance Company for Insuring Houses for Loss by Fire, the second company in America, was popularly known as the "Green Tree" after its leaden fire mark of a tree, signifying the company would insure homes surrounded by trees. The Philadelphia Contributionship would not, claiming that trees attracted lightning and prevented fire apparatus from getting close to the fire. Fire insurance today will pay only if the damage is caused by an "unfriendly" fire and if the insured attempts to limit the loss on the property. Goods are insured for their worth, less depreciation.

Today, about twelve thousand lives and billions of dollars are lost annually by fire. Fires are not a remote possibility; they can and do occur. Some fire fighters

A collapsing wall can topple suddenly and unexpectedly, injuring or burying men and equipment working on the street below. Though the water tower had a strong stream on the briskly burning fire, pressure from within the building caused the wall to explode. Milwaukee Fire Department.

are of the opinion that little is done to prevent fires and to protect against damage from them until a major tragedy occurs. Fire is more often regarded as an aid, used as our friend, than looked upon as our enemy. Yet when it does occur, it does not occur by its own choice. It becomes a hungry monster first feeding on our neglect and then on our property and maybe those we love. It teaches us lessons which are quickly buried and forgotten until the next major blaze. Our memory is short and our reaction short-sighted.

In any case, when a blaze occurs today, we are confident that the fire departments can handle it. They have always been there when we have called them, and they are thought to have such advanced technical knowledge and such advanced equipment that no task is impossible. Many large departments

require vocational training with courses ranging from Fire Fighting Tactics and Strategy to the Psychology of Human Relations. But even with their knowledge and equipment sometimes life is lost. In December 1966, twelve fire fighters were killed in New York when a floor collapsed and sent them plunging into a basement inferno. Three firemen died January 10, 1973, when the roof of a Chicago cafeteria collapsed on them while fighting a fire. Many professional fire fighters have perished in the line of duty through the years beneath collapsing walls, under avalanches of rubbish, or by heart attacks resulting from the demands of a big fire.

Partially through the efforts of these men, civilian losses are no longer what they once were: 250 dead in the Chicago fire of 1871 or the nearly 1,500 who perished in the Wisconsin and Michigan forest fires of

1871, or the 491 victims in a Boston nightclub fire in 1942. After all, we no longer see men dashing through narrow streets pulling apparatus which had to be filled by a bucket brigade. We no longer see the horse-drawn steamers, once "the most efficient apparatus of the day," attempting to get their weak streams on the fire. Today the equipment is shiny, scientific, modern. The fire fighters are highly skilled and dedicated. We have ladders that can reach 125 feet and pumpers capable of 2,000 gallons per minute. There is chemical equipment for stubborn fires and rescue squads equipped to save lives. And the fire chief's trumpet through which he barked orders ("Bring 'em closer, bring 'em closer!" and "Roll 'er out!") has been replaced with modern two-way radio communication.

Today the feeling is that large gatherings of people are less likely to panic. But the danger is still there. Collinwood, Ohio, 1908: 176 children perished in a school that held fire drills. Chicago, 1958: Our Lady of the Angels school burned. In the ashes 90 bodies of children were found.

How much can we take for granted? Today's fire fighters do not carry miracles in their shiny, red trucks. The fire axe is not a magic wand.

"We sell service," said one former battalion chief. At one time, the service did not need to be sold. It was enough that the fire department was there, putting out fires, doing its job.

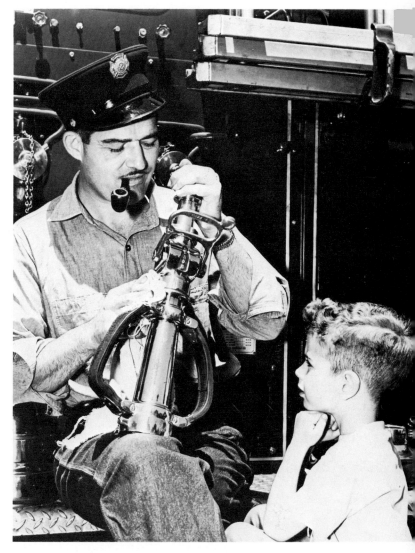

A boy's traditional hero is the fire fighter since it is he who risks his life daily for others and who tells so many interesting stories about his exciting life. Here, a boy named Tommy, much too young to join the department, is learning a fine point of fire equipment. State Historical Society of Wisconsin.

Who is the fire fighter?

The person who responds to society's needs with compassion.

The person who can get a cat out of a tree.

The person who has chosen "the world's most dangerous occupation" as a livelihood.

The person who can save another's life in a physical sense, as others save in an emotional, psychological or social sense.

The person who bodily protects life and property in a country that honors intellectual achievements.

The person who still works fifty-five hours, or more, a week.

The person who has hundreds of stories to tell about the big ones.

The person that a little boy can really talk to.

The person that people delight in calling to the address of a non-existent fire. About thirty-five percent of some companies' calls are false alarms.

The person who sometimes is greeted with bullets when he's just doing his job.

The person who, on attempting to extinguish an arsonist's blaze in an abandoned building, finds the stair treads cleverly removed, and drops to possible serious injury or death.

The person who enters a burning building and, risking his own life in the dense smoke, searches the rooms for trapped life.

The person who rescues a puppy and massages his heart for twenty minutes to bring him back to life.

The person we forget all about until we smell smoke.

To this fire fighter this book is dedicated.

Early America in Ashes

There were two fears that headed the list of frustrations of the first colonists in America. The list included such dreaded calamities as epidemic diseases and inevitable starvation, but what caused the most concern were Indian attacks and fires.

The colonists were quick to observe that the Indians were fighters. The Indians resented the intruding white colonists and, often, were not even kindly to other Indian tribes. William Bradford, Governor of Plymouth Colony, wrote that Indians "delight to torment men in the most bloodie manner that may be; fleaing some alive with the shells of fishes cutting of [off] the members, and joynts of others by peesemeale and broiling on the coles, eate the collops of their flesh in their sight whilst they live, with other cruelties horrible to be related." How much of Bradford's account was fact and how much was imagination is not certain, though the colonists knew Indian attacks to be extremely destructive, despite primitive Indian weapons that consisted primarily of arrows and knives. Throughout early American colonization, care was taken that Indians never received firearms. A few colonists accused and convicted of supplying the Indians with arms were banished from the country.

In 1605 the English adventurer, Captain John Smith, attempted to colonize Virginia, the New World territory named after the "Virgin Queen," Elizabeth. She had tried unsuccessfully for years to colonize the area. James I of England was her successor and retained, if nothing else, Elizabeth's enthusiasm for establishing a settlement in Virginia. Jamestown, the first permanent English settlement in America, was founded in 1607. The Indians were unhappy with the situation, and the settlers were perpetually loading muskets and water buckets to ward off attacks and fires. That same year the Indians captured John Smith, and the Indian maiden, Pocahontas, saved his life in a well-known story.

Fire fighting strategy was primitive at best in this New York fire in the 1730's. All able-bodied men were expected to answer the alarm and men can be seen rushing — and tripping — in the excitement. Some hauled buckets of water to keep the engine's cistern filled, while others pumped the engine in a see-saw fashion to force a stream (or trickle) onto the blaze. Still others empty the burning building of its contents in pessimistic anticipation of the engine's failure.

While Smith was still in captivity on January 7, 1608, a great fire occurred at Jamestown that, according to Captain Smith, who was released shortly afterwards, destroyed "most of our apparell, lodging and private provision." The settlers depended entirely on England for supplies and this incident was too much for many that had clung for so long to so little. Smith's account continues: "Many of our old men diseased, and of our new for want of lodging perished."

Early American settlements were built very much alike. They were, first of all, built quickly since the people walking from the ship had no shelter whatsoever. Holes dug in the sides of hills were all right as temporary homes while trade with the native inhabitants of the land was negotiated. But these people could not adjust to burrowing into the land and calling it home. Soon little wooden dwellings with highly combustible thatched roofs were erected. There were no building codes; they seemed to be impractical at a time when hunger, disease and Indian arrows were taking their toll. Jamestown population decreased from one hundred hopeful settlers in 1607 to thirty-eight physically weak and diseased survivors one year later. Life was not easy in any settlement in the New World, and Jamestown, having no precedent and being surrounded by unfriendly Indian tribes, probably had it worse than the others that followed. In addition to the sufferings of hunger and disease, a historian commented that "as if to aggravate their sufferings, a fire consumed their dwellings in the depth of a severe winter, and left them exposed to the life-chilling blast. . . ." Fire was particularly devastating in the settlements since the homes were built close together to protect the settlers from outside foes. Also, dwellings were not built in logical patterns on neat, intersecting streets, but rather wherever it was practical and thought necessary to build them.

Chimneys were made fire resistive by plastering mud and clay on the wood, the primary element in home chimneys. This presumably kept sparks from igniting the wood which inevitably would burn the thatch on the roof. Unfortunately, the mud dried out and cracked, exposing the flammable wood beneath. Alert observers were quick to report that nearly all the fires in the settlements of the area known as the Massachusetts Bay Colony were caused by faulty chimneys. Officials grew tired of the costly fires, and the Deputy Governor of Massachusetts Bay Colony, Thomas Dudley, proclaimed in 1631 that "noe man shall build his chimney with wood nor cover his house with thatch." The ruling was not heavily protested since colonists were beginning to replace some of the wooden dwellings with brick ones.

Reverend Francis Higginson was quoted as saying in 1630:

> It is thought here is good Clay to make Bricke and Tyles and Erthen-Pots as needs to be. At this instant we are setting a Bricke-Kill in worke to make Brickes and Tyles for the building of our Houses.

Nevertheless, the primary building material was wood, and Jamestown, though resistive to new building innovations, was not resistive to fire. The town burned several times and though a historical account of the town said that after a fire "everyone went steadily to work to repair the damages occasioned by the fire, and the town soon arose from the ashes," this was done only so many times before its inhabitants grew weary. After Jamestown was burned to the ground in 1676, the town began to decline until its existence was history.

While Jamestown struggled for survival, the Pilgrims landed at Plymouth in 1620. On November 1, 1623, a fire swept through seven dwellings. The future of the town was doubtful afterwards since nearly all its provisions were gone. Governor William Bradford noted that the fire began next to a storehouse "in which were their commone store & all their provissions: ye which if it had been lost, ye plantation had been overthrowne." Bradford was not shy about taking the credit for quenching the fire. He continued, "But through Gods mercie it was saved by ye great dilligence of ye people, & care of ye Govr & some aboute him." The conservative Puritan not only blamed the fire on "some of ye sea-men that were roystering in a house wher it first begane" but indicated that goods were not thrown from stores during the fire since it was his conviction that "ther would much have been stolen by the rude company yt belonged to these 2. ships." Instead the fire was fought at the storehouse with wet blankets and "other means." The fire began and adhered to the usual colonial pattern. It first "broke out of ye chimney into ye thatch," proceeding from there to burn the houses and the goods within them.

The first settlement on Manhattan was due to fire. Dutch seamen patrolled the area with a steady lookout for trade with the friendly Indians there. During the winter of 1613 Captain Adriaen Block and his Dutch crew were conducting a profitable trade with the Indians at the mouth of the Hudson River. Their single idea was to trade profitably and then hustle home with their fortunes. This plan was abandoned, however, when their ship caught fire one cold winter's night and was completely consumed. The rest of the winter, of necessity, was spent on

Members of the Rattle Watch (left) were costumed in long robes as they watched for fires in New Amsterdam in the mid-1600's. The colonial fire fighter (right), though simply attired, had more advanced equipment following the invention of hose in 1672.

shore. The Indians helped the sailors to survive the winter, and the crew was able to leave in the spring. Upon their return home, they urged Holland to reap the benefits of trade in the New World. The enthusiasm of the crew pushed a trading company to establish itself in what was called New Amsterdam. Soon after, in 1628, a home caught fire, burning to the ground. This was New Amsterdam's first recorded fire. During Governor Peter Stuyvesant's reign the first building code was established. He saw that a committee was formed whose duties were, among other things, to inspect the chimneys of New Amsterdam. If any were found faulty, the committee had the privilege of fining their owners.

In 1658 New Amsterdam, under Stuyvesant's leadership — or more accurately, under Stuyvesant's orders — purchased a number of buckets to use during fires. Also in that year New Amsterdam's first police protection and fire alarm was established with the Rattle Watch, so named because of the loud noisemakers carried with the members on their rounds to look for fires. However, the watchers were given too much authority and were not trusted by the people. Even Stuyvesant, who had given himself much authority, was not trusted by the people. In 1664 he tried to rouse his citizens to alarm claiming that the British fleet then anchored in the harbor planned to conquer the city. The people were unaroused and did not resist the British, who easily won the town and renamed it New York. New York's history shows that it established fire laws early. Consequently, New York did not have any serious fires until the so-called Negro Plot of 1741. Boston, however, while comparable in size, had four major fires in the same period.

Boston's records show that in 1631 Thomas Sharp's house caught fire. Sharp's home was equipped with the traditional thatch roof and soon this was brightly ablaze. The wind delivered the fire to the Coulburn home and that, too, was burned.

Boston had the habit of stockpiling its ammunition and in 1645 eighteen barrels of gunpowder suddenly exploded. Several buildings and many important records were destroyed by the ensuing blaze. City officials decided that it was folly to store so much gunpowder in one place. Inhabitants of Boston soon found themselves each the owner of one barrel of gunpowder to be stored wherever convenience would permit within the home. So in 1653 Bostonians observed a big fire that was consistently made bigger by the explosion of gunpowder in each home that the fire touched.

The fire took place on a cold, unamusing night in the winter and was billed as the first of the great Boston fires. The city's population was sound asleep when someone, upon observing a fire near the waterfront, dashed from his bed, dressed hurriedly and excitedly began to arouse his neighbors. A brewery, wharf and warehouse were lost as the townspeople threw on their clothes and ran to the harbor, only to watch helplessly. They had no equipment and no voice of authority to organize and help put out the fire. A tavern and other buildings were consumed as they watched. Individual fire-fighting efforts consisted of placing wet blankets on properties.

Town officials, upon studying the situation, determined that a fire stop could be made. Houses and barns only guessed to be in the fire's path were torn down. It was not difficult to pull the flimsy structures down; a simple tug with a large hook was often all that was needed. The fire continued and fed on a barrel of gunpowder in a tavern, blowing it up in a shower of sparks. A few sparks found their way ahead of the main fire to Reverend John Wilson's home. Desperate attempts were made to save it, and water was being thrown on the structure when somebody suddenly remembered the gunpowder. The effort to save the good Reverend's home was abruptly abandoned, and just in time. The gunpowder exploded with a roar and Boston now found itself fighting two fires. The two fires moved toward each other and when they finally merged, they were too weak to do any more damage. One-third of Boston was completely destroyed, including many warehouses that stored supplies. At least three deaths were reported, and these were three little children asleep in the commotion who could not escape when the fire engulfed their home.

Strict fire laws were later passed, including laws specifying that each home have a ladder, a twelve-foot pole with a swab at the end of it for roof fires, hooks, and a provision was made for a bellman to watch for fires. A year later Boston owned some sort of primitive fire engine to carry water to a blaze.

A few years afterward, Boston was plagued by a series of arsonist blazes. A serious fire in 1676, since its cause was never ascertained, was speculated to be of incendiary origin. This fire was far more destructive than the one in 1653, destroying fifty homes, warehouses and stores, and the church belonging to the alarmist American clergyman, Increase Mather. Then in 1679 an arsonist set fire to a tavern sign. The fire, as the firebug probably anticipated, did not confine itself to the sign and proceeded to destroy seventy warehouses and eighty homes. The town conducted a campaign to rid itself of the firebug menace. The threat of execution was made public. Some suspected arsonists were banished, while the remaining population had to give an oath of allegiance every three months. The arsonists, however, still managed to get away with their crimes most of the time.

Philadelphia, the country's second largest city after Boston, had a large fire in 1730, just seven years after Benjamin Franklin came to the city. The fire burned buildings and storage houses at Fishbourne's wharf.

Fires in early America always drew crowds but rarely enough water to extinguish them. Benjamin Franklin recognized the importance of fire protection and, in 1736, organized the Union Fire Company, America's first organized volunteer fire company. In 1752, Franklin founded the first successful American fire insurance company. And the romance of fires and fire fighting had begun.

Benjamin Franklin formed the Union Fire Company in 1736 in Philadelphia. This portrait was painted by C.O. Wright in 1795, five years after Franklin's death. State Historical Society of Wisconsin.

Nineteenth-century printmakers Currier and Ives published a popular series of prints called "Life of a Fireman" in 1854. The artist, Louis Maurer, regarded this series as his finest. Here, a hand pumper is being pulled on "The Night Alarm" as the chief cries, "Start her lively, boys!" Insurance Company of North America.

From Political Revolution to Industrial Revolution

Boston, compared to other cities, burned frequently; another fire devastated the town in 1760. This one, however, was enough to set the city's growth back, and Philadelphia replaced Boston as the colonies' largest city. Apparently, the Joseph Jynks engine purchased from London in 1679 had certain inefficiencies and limitations that could not overcome Boston's numerous fires. New York, too, was growing rapidly and was in strong contention for the country's second largest city. Boston citizens, frustrated and fire-weary, petitioned an arrogant British Parliament for aid after the 1760 fire. Rebuilding after so many fires had taught Boston residents perseverence and certainly patience. Yet even the patient people of Boston were incensed when, two long years later, the only word they received was that King George III, coronated seven months after the Boston fire, had received their request. Boston found that its best friends were the other colonies, who generously gave aid and further widened the chasm between America and England.

What had once been tiny sparks of revolution increased to glowing embers. Many people arguing in favor of separation from England also took an active interest in local affairs, particularly fire protection. John Hancock and Samuel Adams, two of the outstanding Revolutionary personalities, were appointed as firewards of Boston in 1766, and served for nine years. A fireward was the main authority at the scene of a fire, a forerunner of our modern fire chief, although he did not receive compensation. It was his duty to prevent looting, maintain discipline among the fire crew, and direct all equipment.

Anti-British sentiment grew, and the Tory Governor of Massachusetts became uneasy. He sent dramatic requests to England for military aid. When the troops arrived in Boston, American resentment greatly increased — the British were hated, not respected.

33

Whenever a fire occurred in early America, the church bells would ring out the alarm with short taps of their clappers. At 9:15 p.m. on March 5, 1770, Boston church bells clanged with the short, nervous taps that indicated a fire alarm. Only this time there was no fire, and the colonists who turned out for the "fire" suspected there was none. It was thought that the British had turned in the alarm, probably to demonstrate exactly who had the upper hand, so most of the colonists were not armed with buckets but rather with clubs. They arrived on the scene of the nonexistent fire and began to taunt and jeer the hated British soldiers. The Boston crowds continued to linger and the British soldiers formed a line. Still the colonists poured forth their insults until the situation became too much for even the disciplined soldiers. The troops fired into the crowd, killing five colonists. What the church bells had proclaimed to be a fire went into history as the Boston Massacre.

As hostile feelings toward the British continued to grow, the business of everyday life could not be set aside. Records show that John Hancock purchased a new fire engine for Boston in 1772.

The type of engine in use at the time was a pumper-type first designed by Richard Newsham of London. An advertising circular described them as

"substantial" — certainly not an exaggeration. Its power was provided by men — several of them — who pulled the clumsy apparatus to every fire with determination, dedication and just plain muscle. Since there were no organized paid fire departments at that time, the response to an alarm stemmed from a feeling of civic pride. In addition, there was romance and excitement, and the operation of the machines. Firemen have always had an affectionate feeling toward their fire engines, much like the special feeling a captain reserves for his ship. Love is blind, and it must have taken a real love to tote these machines to a fire. They were huge and heavy — twenty men were often needed to operate them.

On December 3, 1731, some of these engines arrived in New York and an account of the newsworthy event stated that it took forty volunteers to pull them from the docks to City Hall. Both of the machines were thirteen feet long, but their ability to save New York from fire came to be seriously doubted. At their first fire, the house burned to the ground. The Newsham engines had long wooden bars on the sides, called brakes, which were moved up and down by the men. This motion touched a metal beam that motivated the piston pump. Newsham thoughtfully invented a suction hose which could fill the engine from a well or cistern. If wells or cisterns were not available, there were always the ancient, but dependable bucket brigades.

It did not take the Americans long to improve on the Newsham engines and make their own. In 1743 Thomas Lote of New York built the first successful American engine. The brakes on the Lote machine were located on the ends, rather than the sides. Since he was a coppersmith by trade, he used extensive brass in the fittings and one of his engines was appropriately dubbed "Old Brass Backs." The engine that John Hancock purchased for Boston in 1772 was probably of this type.

In 1773 the Boston Tea Party occurred and prompted King George to send the British army to that city. The American rebels, pending the arrival of the British troops, chose to leave the city, and the population of Boston decreased sharply. British General Gage suspected that with most of the rebels in hiding outside the city, a plot might be hatched to burn Boston. The general's suspicions were justified; shortly after the troops arrived a fire occurred. It cost twenty thousand pounds damage and destroyed

George Washington maintained an interest in the fire service all his life.
As a member of the Friendship Volunteer Fire Company of Alexandria, Virginia,
he donated this engine in 1775. Insurance Company of North America.

thirty buildings. The colonists still residing in the town were extremely dismayed to learn that the soldiers were not at all knowledgeable in the operation of Boston's ten fire engines. Boston citizens yearned for their regular volunteers and firewards; the fire would have been quenched much sooner had they been there.

Other burnings took place in the duration of the Revolutionary War. During the Battle of Bunker Hill alone it is reported that nearly four hundred buildings were set on fire. The winter of 1775-1776 was bitterly cold and many Boston buildings were either chopped or burned to provide warmth for the unwelcome redcoats. The American rebels were not above using fire for military advantage and burned a house which stored British ammunition. They later set fire to a lighthouse that was of great strategic value to the British army. Tired of the aggravations, the British left Boston. Bostonians again returned and repaired their beloved engines; Paul Revere began serving a four-year term as fireward.

George Washington was probably the most famous fireman of the period. In his youth, he served as a volunteer fireman in Alexandria, Virginia, and was known throughout his distinguished military and political career to often stop and visit local fire companies. Washington was later made a member of the Friendship Fire Company of Alexandria and even bought an engine for it.

New York City was a prized strategic center in the Revolution and on August 22, 1776, the British army anchored in its harbor. Many in the city were loyal to England, and General Washington thought that a military advantage could be achieved by burning the city, reasoning that no one would want to occupy a burned-out city. The Continental Congress, however, did not see it that way and vetoed the proposal. On September 16, 1776, British soldiers calmly occupied New York but on September 21, by more than a simple coincidence, the city experienced the largest fire in its history to that time. There is little doubt that it was a rebel act of sabotage. Numerous fires broke out in various sections of the city, mysteriously all at the same time — 1 a.m.

The arsonists were fortunate to have the aid of a breeze which nudged the flames from building to building. Occupants and the troops ran to the belfries to clang out the code for "Fire!" They were quite

General Ross's British troops burned Washington, D. C., in August 1814. The foreground shows the destruction of the flotilla, while the flaming building in center background, just below the British troops on the embankment, is the President's House. Library of Congress.

disappointed to find the bell towers empty: The ingenious rebels had thought to take the bells away. In addition to delaying any fire alarm, the metal, when melted, would make excellent ammunition. The colonists had also remembered to leave the city's fire engines in a non-operating condition. As the alarm was being shouted through cupped hands all over the city, the fire progressed nicely, unchecked. Some grabbed buckets, but the bottoms had somehow disappeared. The British became more and more irritated, and anyone suspected of the crime of arson found himself at the end of a bayonet or thrown in the flames. A wind change and a stand made with the few uncrippled buckets saved New York from burning to the ground. Almost five hundred buildings, one fourth of the town, were destroyed and the victimized troops had to erect tents for shelter.

Firemen in New York and Boston were exempt from military service during the war. The people had come to respect fire-fighting ability and, after the truce was signed, people in towns and cities were becoming more dependent on organized fire fighting. And toward the turn of the eighteenth century America was vigorously accepting the challenges that

lay before it. The pace of life began to quicken as the country expanded.

In 1785 a new type of fire engine was developed called the "gooseneck" style, and was first popular in New York City. It received its nickname from the long crook that extended above the air chamber case over the main body of the engine. This was the first distinctly American engine and a great improvement over the previous type. It was not as heavy and bulky as the Newsham machine, yet it had a larger capacity. Also, the position of the water stream could be changed without moving the whole engine.

Fire companies continued to grow and their engines took on names that epitomized the times. Engines were known to their loyal companions by names such as "Washington," "Congress," "United States" and "Eagle." The men who cared for and prized these engines — the volunteer firemen — were invariably prominent, active, public-minded citizens. Fire companies had limited membership and adopted strict rules. Men were carefully investigated before they were admitted as members. Still there was always a waiting list. Members were fined for failure to attend a fire, failure to wear badges and fire hats,

and failure to attend regular meetings. (Most, however, looked forward to the regular meetings. In addition to official business, there was a chance to socialize, swap stories and tend to the fascinating business of maintaining the engines. Often picnics and sumptuous feasts were prepared.) Fines were also specified for discussion of politics, use of profanity, smoking and being inebriated. Despite the stringent laws, each fireman considered himself very fortunate to be included on the membership roll of a fire company, since he knew that there were scores of others envious of his position.

Serious rivalry between companies developed. It ranged from determining which company could plan the best picnic outing to which company could hook its hose to a hydrant at a fire and get the first water on the building. Jealousy that led to fights was not uncommon, and the competition of drawing first

ented artists. These panels cost the firemen between three hundred and one thousand dollars for each — and the money came from their own pockets.

Inventiveness was always a part of the fire departments, and the first fireboat, Engine 42 of New York City, splashed into the river in 1809. The idea had to be abandoned for a while: The boat was hand-rowed and hand-pumped, and the task of keeping her going was too much for even the heartiest of fire fighters.

A series of grievances developed with Great Britain. In 1807 British ships began to stop and board American vessels, claiming many on board to be His Majesty's subjects and forcing them into involuntary conscription. The United States felt that its dignity was being violated and resentment began to build. In 1812, war was declared on Great Britain. The Americans wanted to overtake Canada, and it soon became apparent that the northern front would be the decisive theater of operations. The British, however, were somewhat overconfident about their naval strength and decided it would be advantageous to batter the populous Eastern coastal region of the United States. It was hoped that victories here would keep American concentration off Canada.

Numerous strikes occurred throughout the coast and in 1814 General Robert Ross launched a campaign in the Chesapeake Bay area. The British decisively won a victory at the Battle of Bladensburg, Maryland, just east of Washington, D.C. The capital city was so near the battle area that President Madison viewed the battle, and only one-tenth of its population remained to see what would happen next. Upon hearing the news of how the battle was going, Dolley Madison, the First Lady, began to pack. In her hastily snatched possessions she included important papers, a painting of herself and the famous portrait of George Washington by Gilbert Stuart. Residents of the city were leaving, and its army, headed by the bumbling Brigadier General William H. Winder, had already left. The army fought in the Battle of Bladensburg, and remained in that area as General Ross, resting his troops only two hours, turned his regiment toward Washington.

President Madison returned to the capital and promptly left again with certain members of his

water almost came to supercede the importance of extinguishing the fire. Competition took form in other ways, too, as fire companies sought to have the most beautiful fire engines. Artists were commissioned to do panel paintings, framed by intricate carvings and placed over the engine's condensing case. These paintings were done by such famed artists as John Archibald Woodside, Thomas Sully, Charles Peale Polk and numerous other unknown, but tal-

cabinet. Wagons loaded with important records streamed out of the city. When Ross entered Washington on August 24, 1814, he and a group of officers rode past some dwellings and were reportedly fired upon. Angered, and probably provoked by fatigue, they set fire to the houses. Shortly after 8:00 p.m. the British soldiers began to set fire to the public buildings. They burned the Capitol, the President's House, the War Building and the Treasury Building; the only public buildings spared were the Post Office and the Patent Office. The flames glowed in the night as the symbols of the young nation crumbled in the fire. The British were undoubtedly impressed by the speed with which the inhabitants had fled: At the President's House they found a table set for a banquet.

The British effort was directed primarily at public, not private, buildings. A thunderstorm in the late evening quenched the blaze. The next day a rare tornado ripped through the area, uprooting trees and destroying other buildings. The British army suffered serious casualties in an explosion on the docks, possibly implemented by Americans, but the tornado was the decisive factor in causing the army to leave at 9:00 p.m. on August 25.

Both British and American authorities denounced the burning of Washington, though it was congruent with the attempt to make the Americans war-weary. If that was the plan, it failed. Many Americans did not have even a remote interest in Washington, and, to those who did care, this was an outrageous act that demanded retaliation.

The British prepared for an even larger siege deeper in the Chesapeake Bay area. Baltimore, the fourth largest city in the United States, was the next target. Acutely aware of the destruction in Washington, Baltimore determined to fight back. Trenches were dug around the city and additional fortifications were built. The battle that inspired "The Star-Spangled Banner" was won by the Americans, and the Treaty of Ghent, signed in Europe on December 14, 1814, ended the war. The treaty restored all original territory and the United States began to turn its attention to internal expansion.

Though much of the nation was moving westward, Eastern cities continued to grow. New York City developed a heavy and healthy interest in trade. It was not welcomed by all, however. One clergyman believed that a calamity would be the price for "the inordinate spirit of money-making" that was a "prevalent sin in our community." A catastrophe that struck New York, particularly the financial center, proved the preacher's theory to his satisfaction.

"Proof" was offered on the winter night of December 16, 1835. One diarist of the period recorded that the temperature was zero degrees. The Hudson and East rivers were frozen and the cold spell had lasted for so long that even peddlers refused to go out on their rounds. Fire fighters were exhausted from two serious fires that had occurred the previous day. One of those fires destroyed nine buildings while the other fire started several more fires, with which firemen bravely contended. In fact, fire fighters might have been worn out from a hectic year: They had been called upon to fight nearly five hundred fires in 1835. People were edgy and bell towers at city hall and the fire stations stood ready to spread any alarm.

Peter Holmes, a private insurance watchman, saw a puff of smoke coming out of a five-story building at 25 Merchant Street, occupied by an importer and some dry goods merchants. Holmes was fearful and frantic and ran around the streets crying "Fire!" The firemen responded immediately and were on the snow-covered streets, dragging their apparatus, as the bells clanged furiously from the towers. Though tired and hampered by the snow which covered the thoroughfares, one engine company arrived within ten minutes after the alarm was shouted at 9:00 p.m. The fire made rapid progress and after burning for fifteen minutes had managed to involve fifty other buildings. The entire department, which included forty-nine engines, five hose carts and six hook-and-ladder trucks, had to be called out. By midnight the fire was eagerly burning a thirteen-acre area; banks, insurance offices, stores and warehouses were wholly consumed.

The firemen did all they could to hold the fire in check but their engines could produce no more than

The first American-built steamer (opposite), designed by Paul Hodge in 1841, weighed almost eight tons. Firemen, who felt their way of life was threatened, resisted the steamers, and both hand and steam apparatus were used in competition for a while. In this 1861 print (above) firemen using both hand pumpers and steamers battle a New York night blaze. Insurance Company of North America.

mere trickles of water which hardly had any effect. Since the rivers were frozen, it was difficult to tap the water from there. Some firemen rose to the challenge of the occasion and moved their equipment onto the ice, cut holes and lowered their hose into the water beneath. But it was hard work. Water pressure was weak and the firemen had to get close to the blaze in order to have any effect at all. While many were burned by the hot tongues of flame that lashed out at them, many others were frostbitten just a few feet away. Some firemen poured whiskey in their shoes to keep their feet from freezing; others scoured the ruins looking for blankets to keep them warm. The machines demanded that they constantly be worked. More than one crew discovered that after working long and hard for a stretch of one hour, they were forced to rest, only to come back to their machine to find it frozen and useless. Others had started too large a fire beneath their engine and were forced to ditch the ruined machine. Firemen were coated with ice, water froze in the hose and the fire continued.

The men of Black Joke Engine No. 33 valiantly and cleverly fought the fire from the deck of a ship docked at port. The men of this company were

A conspiracy of fire and ice nearly doomed New York City during the Great Fire of 1835. Firemen fought gallantly but one by one the machines froze. The city's population turned out to watch their town burn. Through perseverance and ingenuity fire stops were made. Chicago Historical Society.

known to be strong and hale, and proved it by working on the fire with their heavy, clumsy equipment for five straight hours without a rest.

Efforts were organized to stop the fire at Wall Street and prevent it from spreading to New York's northern half. Even though all odds were against them, the firemen rallied to contain the fire here. The roof of a building on the northern side of Wall Street caught fire and it looked as though their efforts were at last doomed. But the firemen were determined to do their job and a few risked their lives climbing onto the roof to extinguish the stubborn flames. A cold, fierce wind blew the firemen off their precarious perch, seriously injuring them. Under competent direction, more firemen were able to climb to a point high enough to get a healthy stream onto the new blaze, and, at last, they were victorious.

Throughout the contest of New Yorkers versus the fire, citizens methodically fetched buckets for the bucket brigades and devised their own means of quenching the fire. A restaurant owner contributed vinegar to be poured into the engines and then forced onto the blaze; at least two buildings were known to be saved this way.

The Exchange Place was a forerunner to the New York Stock Exchange of today, and the Post Office was located in the basement of the structure. It was thought to be a fireproof building and residents,

owners and managers of other buildings in the area quickly stored all their valued goods inside. When it was hinted that perhaps the building was not fireproof after all, the goods were hastily reloaded on wagons and carted away. Sailors ordered to the area aided in the salvaging of goods and attemped to save a large marble statue of Alexander Hamilton. They at last gave up on the statue just before the building collapsed, barely escaping with their lives.

A church in the area caught fire. As its steeple lit the New York skyline, an organist within the church began to play a funeral dirge. The building, along with the organist, perished in flames.

The fire was visible from as far away as Philadelphia. In addition to being bright, the blaze was extremely hot — enough to melt metal roofs, sending molten embers sliding toward valiant firemen.

People came from the two nearby tenement districts of the Hook and Five Points to engage in looting. Some were so successful at it that they were caught setting more fires in an attempt to sustain the golden opportunity that the tragedy presented. In the early morning hours of December 17, the idea of making fire stops with explosives was discussed between Mayor Cornelius Lawrence and Chief James "Handsome Jim" Gulick. They decided to go ahead with the idea, and explosives were successfully used to make effective fire stops. By dawn the fire was

The view from Exchange Place the morning after New York's Great Fire of 1835 consisted of smoldering ruins, belongings haphazardly tossed in the streets and handfuls of people inspecting the ruins in bewilderment. Weary firemen are relieved to be directed back to quarters. Chicago Historical Society.

confined to a few blocks. The firemen had "stuck it out," fighting bravely throughout the night, and a few finally began to pull their equipment back to the fire houses. Reinforcements had arrived — Newark had sent six fire engines and hose carts, and the Philadelphia fire fighters had also answered the New York blaze. The Philadelphia firemen had to detrain at one point and tug their ponderous equipment over a six-mile area that did not yet have tracks, but they persevered and arrived in New York.

Business was suspended after the fire. The mayor was soon busy answering thirty-three lawsuits brought against him for using explosives to make fire stops. New Yorkers looked on the ruins of 654 buildings amounting to a loss of between twenty and twenty-two million dollars, less than half of which was insured. After the fire, loans were recklessly made to businesses for rebuilding. The fire of 1835 was one of the contributing factors to the Panic of 1837, the most serious recession until the Great Depression of the early 1930's.

Chief Gulick received much praise for his work during the fire, except by his political opponents. A series of events removed him from the office of Chief Engineer, which provoked his admiring firemen not to fight any fires on his day of removal, May 3, 1836. The firemen united behind him and later, upon

running for the office of City Register, Gulick won by the largest margin of any candidate up to that time.

The first steam engine was invented by Paul Hodge of New York in 1841. Not only was the pump steam-operated, but the steam moved the self-propelled engine to the fire. It turned out to be impractical and undependable, however, and was later converted to a stationary engine.

It wasn't until 1854 that the engine heralding the "Age of Steam" was invented. Cincinnati, Ohio, councilmen offered a five-thousand-dollar reward to anyone who could build a successful steamer. Alexander B. Latta claimed the honor of the achievement but Abel Shawk provided the engineering brains. They named it the "Uncle Joe Ross" after one of the councilmen. It weighed twenty-two thousand pounds and required four horses, in addition to the steam's propulsion, to get it to the fire. It was at first bitterly opposed by the firemen who felt that their man-and-muscle way of life was threatened. They felt that human brawn was more effective, and certainly more romantic, than a machine that did all the work.

Any company introducing any new innovations in fire equipment knew that the best and only way to get the invention on the market was to travel from city to city, entering the equipment in numerous contests, or musters. Fire fighters, though at first opposed, had to admit that the newfangled machine

Efficient steamers (left) could quickly arrive at the scene of a fire. Upon arrival, a fire fighter's most important function was to save lives, as this Currier and Ives print shows (below). Equipment, dramatic as the fire fighter himself, displayed colorful engine panels, often of such patriotic themes as the signing of the Declaration of Independence (right). All Insurance Company of North America.

could get water on the fire, throwing a stream of water 225 feet.

A new age of fire fighting had begun. It was an age that included the training of fire horses, the wild clomping of feet on the way to a fire, and pedestrians scurrying out of the way. Firemen came to love their horses as well as their smoking, steaming machines. So rather than diminishing, the romance of fire fighting grew.

The country became dynamic, although divided on major issues. The Civil War was the ultimate result and New York was again threatened by a plot to burn the city in 1864, this time by the Confederates. The South was losing the war and it was a last desperate measure. Jacob Thompson, a former U.S. Senator

from Mississippi, headed the plot, which was being directed from St. Catharines, Ontario, just west of Niagara Falls. Messages were relayed between Richmond, Virginia, the Confederate capital, and the Canadian town by a Union double agent who reported all activities to Washington, D.C. New York Fire Chief John Decker had been warned about such activity to take place on election day, November 8, 1864, but Thompson, for some reason, decided to postpone the plot.

A New York chemist had designed portable bombs that combined turpentine, rosin and phosphorus and could conveniently be fitted into a bottle. A total of eight Confederate conspirators were given ten bombs each and registered at several hotels under several aliases on the evening of November 28, 1864.

Shakespeare's *Julius Caesar* was being performed at the Winter Garden Theater when it was discovered to be on fire. The principal actors were three brothers: Edwin Booth, America's favorite actor, Junius Booth, Jr., and John Wilkes Booth, an avowed Confederate who was to assassinate President Lincoln less than five months later. A panic was narrowly averted and firemen, who had extinguished another mysterious blaze at the hotel next door, quickly put out the flames. More mysterious fires broke out at the same time in the St. James Hotel, the Fifth Avenue Hotel and the La Farge House. Chief Decker, alerted to the situation, turned in a general alarm which brought out all of New York's equipment. The Howard Hotel, the Belmont, the Tammany, the United States, Lovejoy's and others were all set afire, and all were found to have the phosphorous bottles. One of the Confederates set fire to Barnum's museum "for the fun of it." But the volunteer firemen were efficient and the plot to burn New York failed. The need for a

INDEPENDENCE,

HOSE 3 COMP'Y

permanent, paid department was realized, and in 1865 New York abolished its volunteer system in favor of a paid department.

Southern cities were also ravaged by fire during the Civil War. Atlanta began as the terminus of the Western and Atlantic Railroad in 1837, and was named after it, Atlanta being the feminine form of Atlantic. Chartered ten years later, it became of vital significance to the Confederacy as a center of manufacturing and supplies. Atlanta's population steadily grew in the 1840's and 1850's though its thoroughfares consisted of red mud, and livestock casually wandering through the streets occasionally caused traffic jams. In 1864 it had a population of over ten thousand, large for its day.

On July 20 of that year, Union forces under the command of General William T. Sherman began a siege of the city, shelling it almost continuously until late August. His army occupied Atlanta on September 2. Although he had promised the citizens that their lives and property would be respected, he nevertheless evacuated the entire city on November 15 and systematically set fire to it. He ordered that those railroad and industrial plants that had somehow survived the shelling be destroyed. But the fire did not confine itself to those commercial centers and four thousand to five thousand homes, churches and businesses also caught fire. Because there was no one to fight the fires, Atlanta was virtually burned to the ground. Sherman's forces left the ruined city behind and continued toward Savannah in the celebrated March to the Sea, burning everything in their way, while Confederate troops reoccupied Atlanta.

State militiamen toured the city after the fire and were appalled by what they saw. Animal carcasses littered the streets and palatial homes were now mere ashes. Trees that once provided shade for a more leisurely way of life were reduced to stumps and twisted iron poked through the rubble, a grim reminder of fire that was used as a weapon in a country torn by war. Few structures survived the conflagration. A priest bravely fought to save his church and several others. A few homes had been spared, presumably because their occupants were Northern sympathizers.

But Atlanta's residents returned, undaunted, to rebuild and reorganize. About fifty families refused to evacuate their homes when ordered to do so, and they greeted the returning citizens. The people responded with enthusiasm. One writer said:

"Let us now look to the future! That which built Atlanta and made it a flourishing city will again restore it, purified, we trust, in many particulars by the fiery ordeal through which it has passed. Soon the whistles of the steam engines will again be heard . . . soon the cars from Macon and Montgomery and Augusta will bear their burdens into and through our city. Ere long, too, we feel confident that the State Road will be in the process of reconstruction. Let no one despond as to the future of our city!"

Within four years, Atlanta became the capital of Georgia and, although another huge fire swept the city in 1917, it prospered to become one of the primary cities in the South.

Richmond had an earlier history. In 1644 Fort Charles was established on the city's site for protection of the nearby James River plantations from the Indians. A town grew up, and in 1779 it was chosen capital of the Commonwealth of Virginia. It became capital of the Confederacy in 1861. Known for its gaity, Richmond was soon stained with the realities of war. Its forty-thousand population was increased by government workers and Confederate soldiers. Throughout the war it was a primary objective of the Union forces; many battles were fought to protect it.

On April 2, 1865, its defenders abandoned their capital city, setting fire to munitions stores, warehouses and businesses, and then burning bridges behind them in front of the advancing Union army. Flames quickly spread and were fed by whiskey which leaked onto the streets from barrels that had been ordered destroyed. Firemen were unable to fight the fires since the people were riot-stirred and attempted, often successfully, to cut the firemen's hose. Ironically, the invading Northern army helped put the fires out. The conflagration destroyed nearly all of the central city, but Richmond's residents, like those of Atlanta, returned to rebuild, and the physical scars of war were quickly eradicated.

Fourth of July celebrations have always allowed people to express themselves in loud merriment. An account of a Fourth of July celebration in Portland, Maine, in 1866 said: "Never had Portland looked more beautiful than when the sunrise-gun boomed across the water, announcing the ninetieth anniversary of our independence. . . . Of course, the popular satisfaction expressed itself in the report of pistols, guns, and firecrackers; and all through the day the usual amusements went on, and in the afternoon almost everybody was on the street."

A high southwest wind blew above the happy scene, and the city was ready for a good fire. Conditions were right since people were scattered and unprepared for a major outbreak and the city, of

Firemen pose in front of a steam engine in the year 1870. Though their posture is casual, their dress is uniform, an innovation that was popular as departments became professional. James Haight.

course, made out of wood, was dry and sun-heated. A fire alarm was turned in around 5:00 p.m. The fire reportedly began when a boy, who wanted to scare some workmen, threw a firecracker into a pile of wood shavings next to a boat-builder's shop. If scaring was the intent, it was an unparalleled success. In an area a mile long and a half-mile wide, 1,500 buildings were consumed. The firemen fought valiantly and, according to one account, they "put forth a strength almost super-human." When it was over, half of the city lay in ruins, with a loss of ten million dollars. Portland quickly recovered, however, and by 1900 had a population of over fifty thousand people.

Throughout the century, Boston continued to burn. A fire in 1824 destroyed sixteen buildings at a loss of $150,000, and a fire in 1825 destroyed fifty stores at a cost of a million dollars. That same year a fire destroyed another ten buildings and in 1835 a fire left one hundred families homeless. It was no wonder that in 1852 Boston was the first city to install a telegraph alarm system, which tapped out an alarm from a fire box near the location of a fire to department headquarters. Church bells could also be used as an aid in determining the location of a fire. If the bells, for example, tapped out two claps, then paused, then banged out five short taps, then paused, then three short taps, the location of the fire could be determined as being in the vicinity of firebox 253, which was located at some particular intersection,

known to the firemen. Boston, perhaps more than any other American city, felt the need to be innovative as far as fire protection went, since it burned so frequently.

Boston's greatest blaze took place on Saturday, November 9, 1872, a little more than a year after the Chicago fire. The department was heavily staffed with 475 members, but its horses were ill with the "epizootic," commonly known as the flu. When the alarm was turned in, the firemen made a fast decision as to whether or not to arouse their horses, some of which had recovered. They decided that the horses' health was too precarious and so they tugged the apparatus through the narrow streets themselves. In addition to the extra time needed to pull the apparatus themselves, the alarm for some reason had been delayed and the fire was quite large by the time the firemen arrived on the scene. They fought valiantly to contain it but the fire was too extensive. At last they were able to make a stand on some land which had been cleared for the building of Fort Hill, and by Sunday afternoon the fire was out. The city had experienced fires before, but none like this — 776 buildings were destroyed and fourteen lives had been lost, half of them firemen.

Improvements continued to be made in the fire-fighting system, and the romantic appeal, heroism and bravery increased as the fires got bigger.

Conflagration in Chicago

Drought dried the Western prairies and north woods in the summer of 1871. Forest fires were frequent in the great Northern forests of Minnesota, Wisconsin and northern Michigan. Farther to the south, in the closely built wooden city of Chicago, the fourth largest city in the country, dwellings, sidewalks and business places were as dry as tinder. The last appreciable rainfall had occurred in July and that measured only two and a half inches. It was October and already the fallen leaves blew in the strong wind; lack of rain had brought an early autumn. Members of the fire department, whose job it was to be concerned with the consequences of a dry, brittle city coupled with a high, gusty wind, were anxious. On the evening of October 8, they were also tired.

The previous night the entire department had been called out to fight a blaze in the West Division, the area west of the South Branch of the Chicago River. It was the worst fire the city of Chicago had known. The fire rampaged over a four-block area and did $700,000 in property damage. The total fire-fighting force of Chicago numbered 185; thirty were injured in that blaze and everyone that fought hard to contain it was exhausted. The fire department owned seventeen steamers, twenty-three hose carts and four hook-and-ladder trucks — not enough for thriving Chicago, the fastest growing city in the world. Some of the equipment was damaged by the blaze on October 7; three of the valuable steamers were rendered useless by falling walls. It was hoped that they could be repaired in a few days. It was also hoped that this would be enough time. It was frightening to think about what would happen should a fire break out that was even near the size of the one the previous night.

Chicago was a city that lived and built under the banner of "hurry." Forty short years previous, in 1830, it was a small fort on the Western prairie with a population of 170. Its location on Lake Michigan and its natural waterway inland promised growth, and by 1840 the population had increased dramatically to 4,800. When the locomotive's whistle was heard in Chicago in 1848, the value of the village was further enhanced and growth continued at a rapid rate. By 1860, the population was over 110,000 and growing so fast that buildings were being erected at a rate of about seven thousand a year. By 1866, Chicago boasted nearly forty thousand buildings, over ninety-three percent of which were of wooden construction. Between 1866 and 1871 more buildings hurriedly

In 1912 artist Julia Lemos painted this view of the Chicago fire which she had seen before she fled to the nearby prairie. Chicago Historical Society.

went up to accommodate the thriving commerce, bringing the total of Chicago's buildings to sixty thousand. These were the finest buildings to be found anywhere at that time. The city and its culture grew: Fine opera houses, hotels, large stores, churches and magnificent theaters lined the streets. Huge grain elevators were constructed and Chicago had the largest grain market in the world. The city had been chartered in 1837, and by 1871 it had swollen to a population of 334,000.

Chicago had many attractive features. It was the place where Cyrus Hall McCormick built his large reaper factory. It was a city of opportunity, and immigrants poured into it from the East Coast. Dave Kennison, the last survivor of the Boston Tea Party, decided to settle there. Fire Marshal Williams had come to Chicago from Montreal at the age of twenty. Among the city's treasures was a bust of Abraham Lincoln, the only one he had ever posed for, and the original copy of the Emancipation Proclamation.

Chicago was not lacking in civic pride and built a fine courthouse on a tract of land in the South Division in the heart of the city. It was built of stone and its imposing block construction was the pride of Chicago. In its vaults were stored all the Cook County records; in its basement were housed all the convicted criminals, and at its crown was a great bell shrouded in a huge dome. Chicago by 1871 had built four thousand miles of railroads, spanned its river with twenty-four bridges, and stored its vast supply of grain in seventeen large elevators.

For all its promise and claim to greatness, there were people and certain businesses alarmed at the careless building conditions in the city. During one year alone, Chicago had seven hundred fires. Lloyd's of London inspected the city and became appalled enough at the conditions there to order that no more insurance be written on any property in Chicago and that outstanding policies be cancelled if possible. Nevertheless, Chicago property increased in value and other insurance companies vied to snatch up the insurance.

After the sixteen-hour battle on October 7, tired firemen left the smoldering ruins in the West Division to get some rest. The fire had begun in an old planing mill. The area was crowded with small wooden houses and the fire had mercilessly eaten through them and then concentrated on the coal and lumber yards in the area. One grain elevator stood tall against the flames and was the only building not turned to ashes.

On Sunday evening, October 8, some Chicagoans were quietly at rest. Others were relaxing and concentrating on a minimum of chores. Happy sounds could be heard in some neighborhoods as immigrants, the new Americans, held small celebrations for whatever special occasion the poor shanties could hold. Others were returning from evening worship. One contemporary described the scene:

> The churches were just dismissing their devout worshippers after evening service, when the fire-bells rang their loud alarum. The evening before, a fire had raged of unparalleled violence, and the embers still glared in the darkness, and people were easily roused to intense alarm. Many hastened from the House of God to the scene of the fire, fearing that the high wind might imperil even larger districts of the city. None dared to dread any such devastation as that which followed.

The bells were bonging out the wrong address and both the crowds and the fire apparatus were hastening to the wrong location. The actual fire was in the poor Irish neighborhood on the corner of DeKoven and Jefferson streets in the West Division. The O'Learys lived there.

Whatever details of the story are missing, it is certain that the fire began in the O'Leary barn. The barn was small, measuring sixteen feet by twenty feet, and of wooden construction. Furthermore, it was stuffed with hay to feed the animals kept there, which reportedly consisted of five cows, a calf and a horse — certainly a cozy, cramped menagerie. For some reason, Kate O'Leary went to the barn that Sunday evening with a kerosene lamp in one hand. Some say she carried a bucket in the other to fetch milk for her tenant who needed it to prepare Tom and Jerry drinks for a party he was giving. Others say she went to the barn to tend to the calf, which was sick, or possibly she went there to milk her cows at an unusually late hour. The lamp she snatched from the house was rumored to be the finest one she owned, complete with filigree work around the glass. If so, she certainly must have been in a hurry.

The story goes on to say that the cow was distressed at being milked at the late hour, which was understandable if the cow had been milked earlier. She set the lamp down on the straw-covered wooden floor and set about to persuade the cow to give her some milk. She was not particularly gentle about the means she used to coax the cow. Kate was getting on in years and had not had an easy life — patience probably left her when she was much younger. The cow, determined to be left alone, kicked backward. She missed Mrs. O'Leary, but hit her best lamp. The kerosene spilled from the lamp onto the straw and briefly threw the scene in darkness. Then the straw burst into flame. Kate O'Leary in her excitement splashed the small amount of milk she had managed to obtain onto the flames. Already the fire was beyond such an attempt and was spreading to the straw in the loft. She ran from the barn screaming, and the whole neighborhood turned its attention to the O'Leary barn.

One neighbor who was at the O'Leary tenant's party estimated the seriousness of the situation and ran to Goll's drug store to turn in an alarm. It would not take long, he reasoned, for all the two- and three-room wooden shanties in the West Division to

catch fire. It didn't take long. Mr. Goll held the key to the fire alarm box and was naturally hesitant to turn in an alarm, particularly when reported by someone so obviously fresh from revelry. But smoke could be seen and the box tapped out the alarm. It was not received. Three other boxes in the neighborhood turned in alarms, but for some reason, none were received until much later.

Engine Company No. 6 saw flames from its watch tower soon after the fire started and roared off to the area of Jefferson and DeKoven. The courthouse watch had also seen the fire but had inaccurately estimated its location, resulting in other equipment ending up more than a mile away from the fire. When Engine Company No. 6 arrived at 8:45, the "Little Giant," an engine put into service in 1860, was immediately put to work on the fire, which by this

time shot fifty- to sixty-foot flames into the air and had engulfed approximately thirty houses.

When Robert A. Williams, the chief fire marshal, arrived on the scene, he immediately recognized the seriousness of the situation, and a second alarm was turned in and shortly after a third. The engines "Waubansia," "Illinois" and "Rice" arrived. "R. A. Williams," the engine named after the chief, and three hose companies were close behind. The wind blew forcefully from the southwest and eddies of wind picked up burning brands and threw them blocks ahead of the fire. Still, it was thought by firemen and spectators alike that the Chicago Fire Department would contain this blaze. They had contained the spectacular fire the night before, hadn't they? They had contained numerous minor fires in years previous, including the rash of fires the last eight days,

not to mention the seven hundred in one recent year. Everyone knew the wells and cisterns were dry and that Chicago was made almost entirely of wood, now tinder dry, including fifty-six miles of wooden pavement and 651 miles of wooden sidewalks. Still the fire brought spectators from all areas of the city, none of them guessing that their own homes might be in danger.

And, despite the gale-force wind, progress was being made. It was then that word was received that St. Paul's Roman Catholic Church, a huge wooden structure several blocks distant, near the South Branch of the Chicago River but also in West Division, had caught fire. Equipment had to be dispatched from the main fire to the new fire, dispelling hopes that either blaze would be controlled. When the firemen arrived at the church the awesome spire was more awesome than ever, having taken on the warm, orange glow of the fire. The four-horse-drawn engine, the "Coventry," was one of the first steamers to get a stream of water on the new fire, but the blaze burned with fierceness and could not be halted. The huge spire crumbled and crashed to the ground.

In Chicago, even the factories were of wood, including the two next to the church. It became evident that if the factories could be saved, the fire could probably be brought under control, but if the factories were allowed to perish and feed the fire, the whole South Division across the river was in great peril. Roelle Furniture Finishing on one side of the church was quickly consumed. The W. B. Bateham shingle mills were just in back of the church, and the 240-foot structure almost instantly burst into flame.

The fire department continued valiantly but as one observer described:

... The flames ran along the wooden sidewalks, and whole tenements would burst into flames as simultaneously as if a regiment of incendiaries were at work. The narrow streets were crowded with appalled spectators, half-dressed women with aprons thrown over their heads running distractedly hither and thither, and men tearing furniture to pieces in the furious haste with which they flung it out of doors or dragged it through the crowd. The element had the best of the battle so far. Engine No. 14, driven back foot by foot, was penned in a narrow alley; in another moment a gush of flame came from the rear, and the firemen could only cover their eyes from the blinding heat and stagger desperately to safety through the burning belt that fringed them round, abandoning the engine . . . .

The two fires — one originating at the O'Leary's, the other at St. Paul's — became one, and flames crossed the South Branch of the Chicago River, entering the South Division. Here just a few short hours before, Chicago's citizens had retired, despite the strange, faint glow in the west. The tolling bells and thunder of fire roused them from their nonchalance. They had to gather whatever they could carry and attempt to escape with their lives. In their panic, they grabbed senselessly for their possessions. Men strapped mattresses to their backs, stupidly

burdening themselves to incapacitation, forced at last to drop the flammable goods into the street for the fire to feed on. Others attempted to remove pianos or other heavy furniture. Rev. E. J. Goodspeed's account describes the scene:

> ... The angry bell tolled out, and in a moment the bridges were choked with a roaring, struggling crowd, through which the engines cleft a difficult way toward the new peril. The wind had piled up a pyramid of rustling flame and smoke into the mid-air.... The earth and sky were fire and flames; the atmosphere was smoke. A perfect hurricane was blowing, and drew the fiery billows with a screech through the narrow alleys between the tall buildings as if it were sucking them through a tube; great sheets of flames literally flapped in the air like sails on shipboard. The sidewalks were all ablaze, and the fire ran along them almost as rapidly as a man could walk.... Showers of sparks, intermingled with blazing brands, were borne aloft by one eddy of the breeze, and rained down into the street by the next, while each glowed a moment and was gone, or burned sullenly, like the glare of an angry eye.... There was fire everywhere, under foot, overhead, around. It ran along tindery roofs, it sent out curling wisps of blue smoke from under eaves, it smashed glass with an angry crackle, and gushed out in a torrent of red and black; it climbed in delicate tracery up the fronts of buildings, licking up with a serpent tongue little bits of woodwork; it burst through roofs with a rattling rush, and hung out towering blood-red signals of victory.

There was no single voice of authority or order. Rev. Goodspeed continues:

> The people were mad. Despite the police — indeed the police were powerless — they crowded upon frail coigns of vantage, as fences, and high sidewalks propped on rotten piles, which fell beneath their weight and hurled them, bruised and bleeding, into the dust. They stumbled over broken furniture and fell, and were trampled under foot. Seized with wild and causeless panics they surged together backwards and forwards in the narrow streets, cursing, threatening, imploring, fighting to get free.

Firemen continued the fight. Much hose was lost from moving the apparatus from one location to another. Some hose burst under the extreme stress; some burned. The courthouse was thought to be the only sanctuary left in Chicago. Surely it would not burn. On one floor of the structure Mayor R. B. Mason desperately tapped out telegraph messages to nearby cities, such as Milwaukee and even as far away as Cincinnati and Dayton, imploring them to send whatever men and equipment they could spare. As the firemen became more exhausted and exasperated, the fire became stronger and more forceful. The flames raged on in the South Division and it appeared that the entire heart of the city, what is now called the Loop, would be lost. Bridges were burning as people jammed the passages across the wide main branch of the Chicago River to the flame-free section to the north.

THE GREAT CHICAGO FIRE

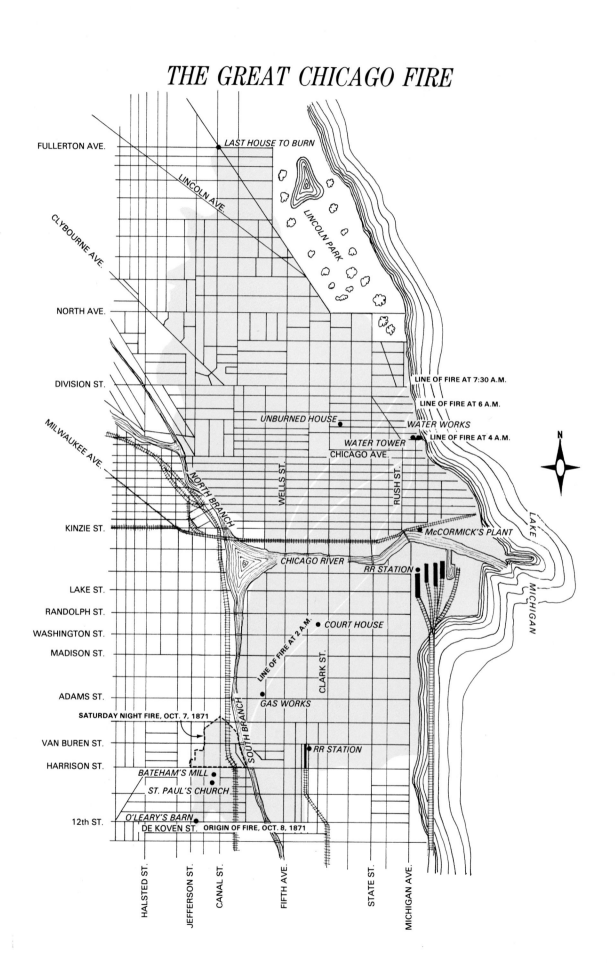

FULLERTON AVE.   • LAST HOUSE TO BURN

LINCOLN AVE.

CLYBOURNE AVE.

LINCOLN PARK

NORTH AVE.

DIVISION ST.

LINE OF FIRE AT 7:30 A.M.

LINE OF FIRE AT 6 A.M.

MILWAUKEE AVE.

UNBURNED HOUSE •

WATER WORKS

WATER TOWER

LINE OF FIRE AT 4 A.M.

CHICAGO AVE.

N

NORTH BRANCH

WELLS ST.

RUSH ST.

KINZIE ST.

McCORMICK'S PLANT

LAKE

CHICAGO RIVER

RR STATION •

MICHIGAN

LAKE ST.

RANDOLPH ST.

WASHINGTON ST.

LINE OF FIRE AT 2 A.M.

• COURT HOUSE

MADISON ST.

CLARK ST.

ADAMS ST.

GAS WORKS

SATURDAY NIGHT FIRE, OCT. 7, 1871

SOUTH BRANCH

VAN BUREN ST.

• RR STATION

HARRISON ST.

BATEHAM'S MILL •

ST. PAUL'S CHURCH

O'LEARY'S BARN

12th ST.

DE KOVEN ST.   ORIGIN OF FIRE, OCT. 8, 1871

HALSTED ST.

JEFFERSON ST.

CANAL ST.

FIFTH AVE.

STATE ST.

MICHIGAN AVE.

Map (opposite) shows the fire's origin in the city's West Division near Jefferson and DeKoven streets. Fanned by strong winds, it crossed the South Branch of the Chicago River and devastated the South Division. Then the conflagration blew over the Chicago River and laid waste to nearly all the North Division. The last building burned at 3:00 a.m. on October 10. An artist imagined this aerial view (above) of Chicago at the height of the fire. Chicago Historical Society.

It was feared that the entire gas works, located in the South Division, would blow up, dooming all. An engineer at the gas works stuck to his post, valiantly turning the valves to let the gas into the reservoir and sewers, despite the flames coming dangerously close to him. Finally, he fled after draining most of the gas. When the fire reached the tanks only a minor explosion ensued, yet the heroic efforts proved to be catastrophic. The crowds trampled each other in the streets with their possessions and families, hailing wagons at the exorbitant rates teamsters were charging. Often the drivers would fling the belongings overboard once they had their money and dash off to search for new victims. Amidst all the confusion, fumes ignited from the sewers, producing numerous small explosions throughout the already terrified, fleeing crowd. In addition to the damage done to the crowd itself, numerous small fires were begun.

Sparks and burning brands had touched the courthouse several times, but fortunately none had taken hold. The windows, however, began to melt and the masonry began to crumble. At last the courthouse could withstand no more and succumbed to the flames. The machinery to toll the bell was set in motion just prior to evacuation of the building, and the bell tolled mournfully and incessantly. The 160 prisoners in the basement were set free and told to

run for their lives. Five other prisoners convicted of murder were handcuffed together and led off by an authority. At 2:05 a.m. the great bell fell to its death, crashing through to the basement, never to ring again. The courthouse, the one building Chicago felt confident of saving, was lost.

The fire began to cross the Chicago River to the North Division. In its path was the waterworks.

Chicago had built the first water tower in the United States. As long as its pumps could provide water, the fight to hold the fire was not entirely lost. The water tower was an engineering feat in its day. The official style of architecture of the building was castellated Gothic, and it contained two stories. All the walls were two feet thick and made of stone. The roof of the main building was constructed of heavy timber and covered with slate, with several holes for ventilation. The boiler rooms were nineteen feet apart and located in the rear of the main building. The floor was of stone also, and since the roof was of iron and slate, no one doubted that the gigantic pumps would keep pumping. The building was, after all, fireproof.

But the wind kept up its fierce howl and hurled large pieces of wood through the air. The ventilators in the roof were open, and sparks seeking new fuel swirled in through the holes and quickly gnawed at

This German lithograph (above) shows the confused scene on the Randolph Street bridge as crowds flee and the fire progresses northward. Chicago Historical Society. The fire razed the courthouse and all of the downtown area (opposite). Wide World.

the huge timber supports. The roof tumbled onto the machinery, and the great pumps ceased to function. There was no longer any hope of staging an effective stand against the conflagration. Pumpers would have to draft water from too far away or draft ineffectively from other hose lines. The fire would have to burn itself out.

McCormick's harvester factory caught fire at seven o'clock the next morning. All bridges over the South Branch of the river were burned; only two were left to take the crowds from the South Division to the North Division.

The people crowded the bridges hoping that the Chicago River would prove to be a barrier against the flames. Many snatched their possessions believing that there was enough time because bridges would hold, but the bridge route of escape as the fire continued its northward rampage proved to be a disastrous mistake for some. An observer of the bridge scene describes:

> But there was one bridge which proved unfaithful to its trust. Chicago avenue bridge appears to have caught fire from sparks before the main fire reached it. Thinking to be able to cross over this bridge, many people delayed their flight, hoping to save at least a part of their furniture before the flames reached their houses. But the delay was too long and the advance of the flames too rapid, and when they finally fled to the bridge it was too late.

It was in flames. Under the approaches to the bridge the exhausted people tried to hide themselves from the flames, the stronger and less exhausted flying to the next bridge north — that at Division street. But the refuge under the bridge soon became a burning furnace. Those gathered under it soon saw the mistake they had made. The despairing ones stolidly stayed where they were, and were suffocated to death. Those with hope still left ran out and attempted to fly north through the flames which were crossing the avenue. A few escaped, but with many it was only a death postponed for the space of a few minutes — burning garments, tottering footsteps, and then a fall to rise no more.

People could not keep their heads in the confusion and many jumped into the river and were drowned. Pregnant women, due to the intense excitement, gave birth prematurely. It was estimated that between four and five hundred babies were born during the fire. Often the births occurred on the street, and often, too, mother and child were burned or trampled to death together.

At 8:00 a.m. on October 9 firemen and three steamers arrived from Milwaukee. They were immediately sent to the North Division. By noon the fire was nearly to Lincoln Park and flames were beginning to heat the exteriors of the homes of Chicago's millionaires.

At 6:00 p.m. a train arrived from Ohio. In addition to three engines from Cincinnati and one from Dayton, the train also carried Miles Greenwood, builder of the first successful American steam engines and chief of the Cincinnati Fire Department, the nation's first paid department. But the fire went on to consume 13,300 homes in the North Division. Only one home stood in the midst of the destruction: a freak oasis, able to withstand such a disastrous test of the elements.

Finally, the fire began to burn itself out on the prairie. It was 3 a.m., almost thirty-one hours after it all began, and a hard rain drenched the ruins.

As daylight broke on the scene of destruction the morning of October 10, it became tragically apparent that nothing had stopped the flames. The burned area from the previous night did not stop the fire in the West Division. In fact, the single grain elevator that had not given in to the previous fire almost instantly gave in to this one. The Chicago River did not stop the flames. At the height of the fire, several buildings were blown up by the army, and this did not stop the fire. The wind had been a determining factor, blowing from the southwest and, together with the fire, creating convection currents that scattered sparks and brands far ahead of the main fire.

About three and a half square miles in the heart of the city, including the entire business section, were destroyed. The Grand Pacific Hotel, the Palmer House, the Post Office, the Chamber of Commerce and six railroad depots were mere ruins. Between 150 and 300 people were dead. The exact number of deaths could not be determined since it was believed that many were wholly consumed by the intense heat. Nearly 100,000 were homeless and either fled to friends and relatives elsewhere or camped on the bleak desolation of city parks or open prairie. Strangely, one of the buildings that did survive was the O'Leary home.

It was reported that all tax records were lost in the courthouse burn. Other valuable county and city documents were lost and valuable business records were now ashes. The Chicago Historical Society was a victim and despite heroic attempts to save the contents, the treasured bust of Abraham Lincoln and the original copy of the Emancipation Proclamation were lost. Almost sixty insurance companies went bankrupt and property loss was estimated at $200 million. Martial law was declared by the mayor. Lieutenant General Philip H. Sheridan, the Civil War hero, and his army troops saw to its enforcement along with Allan Pinkerton's special police who vowed to kill anyone caught looting. General Sheridan opened barracks and set up tents for shelter.

Many were certain that Chicago would never recover. Some who had been a vital part of the city's growth found that it was too much for them to bear to see it crumble. A few left the city. Some sought a scapegoat and blamed the O'Leary cow or the fire marshal. But the majority of Chicagoans vowed to

Thieves & Burglars!

OFFICE OF

Pinkerton's Police.

Orders are hereby given to the Captains, Lieutenants, Sergeants and Men of Pinkerton's Preventive Police, that they are in charge of the Burned District, from Polk Street, from the River to the Lake and to the Chicago River. Any person Stealing or seeking to steal, any of the property in my charge, or attempt to break open the Safes, as the men cannot make arrests at the present time, they shall

Kill the Persons by my orders, no Mercy Shall be shown them, but Death shall be their fate.

Allan Pinkerton.

Looting was a problem after the fire (above left) and strict measures were enforced. General Philip Sheridan's troops eventually enforced the martial law. Chicago's optimistic spirit could not be marred, however, and the Chicago Evening Mail *proclaimed on October 11, 1871, "Chicago will rise again." Rebuilding (above right) began before the ashes cooled. Both Chicago Historical Society.*

rebuild the city and make it even stronger and greater than before. Even more amazing than the pledge was the fact that they succeeded. Three days after the fire, the *Tribune* managed to get an edition in the streets with the headline: "Chicago Shall Rise."

Before the ashes had cooled, temporary businesses had been erected. Aid poured in from all over America and from foreign countries. The *London Punch* said:

> We suppose that the most costly pail of milk ever heard of in the world was the pail which burned Chicago. The gallant Americans are the last people to cry over spilt milk or burned cities.... The Americans remembered us in the time of Ireland's hunger and of the cotton famine, and must now allow us to remember them. And let's be quick about it, or the city will be rebuilt before the money gets there.

Relief centers were set up and they distributed food, money, blankets, clothing and, ironically, non-explosive lamps for the impending winter. The Shelter Committee designed simple, one-room structures for individual families. Within one month, hundreds were being erected. In all, eight thousand were completed. Field, Leiter and Company, the ancestor firm of Marshall Field and Company, completed a new wholesale building within one hundred days after the fire. The ruins were still smoldering when the Chicago Board of Trade met and requested that its building be rebuilt soon. Six weeks after the fire, construction was begun on 212 buildings. Civic leaders did much to encourage the rebuilding efforts. William Bros, president of the *Chicago Tribune*, brought aid from other cities through personal contact. Potter Palmer, who had seen his Palmer House completed just thirteen days before the fire, erected a new one almost immediately. Many of the

The fire did little harm to the water tower but destroyed the pump house, cutting the water supply. Today the John Hancock Center dwarfs the tower. Chicago Convention and Tourism Bureau.

new buildings were outstanding in strength and design and remain today. Among the noted architects who designed them were Louis H. Sullivan, Dankmar Adler and Daniel H. Burnham; they helped establish Chicago's reputation for architectural beauty.

The Chicago fire made history. While some of the people turned into hucksters, patroling the streets claiming to have in their possession — and for sale — the horns or the skin or the tail of the infamous cow that caused all that destruction (someone even made a tour of the country speaking to groups with a real cow and lantern, supposedly *the* cow and lantern), the rest of Chicago was engaged in the task of rebuilding. On the first anniversary of the fire more than half the city was rebuilt. Gala celebrations took place. Chicago was on the road to recovery. Four years later, few traces of the Great Fire remained.

Jonas Hutchinson, a Chicago lawyer, said in a letter to his mother while the fire still burned:

> The whole city is in grief. . . . My office is gone. I am stripped and you may conclude that I am about vanquished. I cannot see any way to get along here. Thirty years of prosperity cannot restore us.

Hutchinson was wrong. The incredible industry and ambition of Chicago's people were able to bring their city to even greater prosperity. Property values, particularly in the business section, increased dramatically after the rebuilding. The population had increased to 500,000 by 1880, and over a million at the turn of the century, attesting to the faith and confidence of the people in a great city — a city that was able to overcome the damage done by one of the most destructive fires in American history.

The Heavens Rained Fire in the North Woods

There was nothing but trees as far as the eye could see in the pine forests of northern Michigan and Wisconsin. City dwellers migrating north to the city of Green Bay, Wisconsin, and from there venturing deeper into the north country were amazed: nothing but trees for hundreds of miles.

A heavy carpet of needles covered the rich, black earth capable of producing tall, living giants. Some white pines were two hundred feet tall. The earth was moist in many places with numerous swamps, bogs, streams and rivers.

The Indians were the first to live in this awesome wilderness. Few trees were cleared away to provide space for their homes. Indians were hunters; often, they burned their game out. When white settlers began trickling into Wisconsin, they found the upper two-thirds of the state covered with forests, with only occasional Indian clearings.

The trees, no matter how majestic, were a nuisance to the hearty settlers. They were simply in the way. From their point of view, trees were an inexhaustible resource, anyway. So the immigrants to this new wilderness — some from Canada, some from Germany, some from the Eastern United States, arriving in the north country with only a dream and the ambition to make it come true — located an axe or two and began boring holes in the vast green carpet that blanketed the upper Great Lakes states. No one considered himself an enemy of the land in the mid-nineteenth century: It was an obligation to cut the trees and enable families to first survive in the wilderness and eventually profit on self-sufficient farms. Sometimes fire was employed to clear the land with an efficiency unobtainable by the slow chopping of the trees with an axe. If the fire burned a little beyond the intended area, no one was much concerned. They thought the trees would last forever.

It was not until later that a few concerned individuals concluded that forests were, unfortunately, not inexhaustible. The holes in the green carpet ripped wider as more land was needed for the industry that the trees provided, for the settlements

that the trees made possible, and for the farms so necessary for the families in the area. Settlers, unthinking and ungrateful for all that the trees provided, eventually came to the realization that the soil which produced the white pines did not provide equally well for farm crops or huge metropolitan areas. (The goal in at least one northern Wisconsin town was to grow and attain a level of prosperity that would rival or surpass Chicago's.) That soil provided best for trees.

The huge pine forests that existed in the region are gone now. The trees in northern Wisconsin today are maple, oak and sumac, with only occasional pines. Their roots extend into the humus-rich soil beneath the tourists that flock there in summer. The countryside is dotted with small towns, usually under ten thousand in population. Many of these towns were built on several inches of sawdust with boards from the sawmills.

The lumber industry reached its height in the 1890's. In 1898 Henry Gannett, the outstanding American geographer, wrote, "The principal enemies of the forest are the axe and fire." Both were used by the first northern Wisconsin settlers to clear the land of its "nuisance" trees. It was an easy step from cutting a clearing to the industrialized carelessness that ravaged the area. Yet this was the foundation for America's lumber industry, the industry that provided wood for an expanding nation.

Lumberjacks were often blamed for carelessness in starting forest fires that often burned out of control over a large area. Concern for the forests ever giving out still did not materialize, though each forest fire was costly to the industry. When an area would catch fire, the spectacle was grand — first, fire in the tree top, dramatically playing with its majestic green crown, highlighting the thick, black clouds of smoke that billowed behind it, and then forcing the tree to the earth with a crash, after torching neighboring trees. Such extravaganzas destroy both the mature trees, ready for profit, and the young growth, the

In 1972 forest fires burned over $21 million of valuable timber and over 117,000 of the 185 million acres protected by the U. S. Forest Service. The 1871 Peshtigo fire destroyed over a million acres. U. S. Forest Service.

future forests.* By the turn of the century concern for the forests was growing. One person commented on forest fires, saying, "It is a magnificent spectacle but one too expensive to be indulged in even by Americans." And whether or not the lumberjacks were rightly or wrongly accused, it can be said with a fair amount of certainty that the rough, rowdy figures rumbling into the quaint little towns on Saturday evenings and Sundays, staggering among the solid citizenry in its best clothes on its way to church, were not always looked upon with a genuine liking or admiration.

---

* However, forest fires are also recognized as being constructive to the future growth of the entire forest when started naturally or under scientifically controlled conditions. But this in no way alleviates man from the responsibility of being careful in the forest.

The north country contained five million acres of forest when the first white settlers arrived. By 1871, half of the forest was gone, much of it wasted. The idea of turning trees to profit had occurred to many. One to whom it occurred was William B. Ogden, a Chicago businessman, who looked at the pines and saw opportunity. He knew that lumber from white pine was much in demand for new buildings. He looked at the sparkling flow of the Peshtigo River, winding its way eastward to Green Bay, the large inlet of water jutting south from the top of Lake Michigan. The river began its brisk stream in the farming region and thick forest to the north and west of the village of Peshtigo. The spot where the river widened at Peshtigo was the spot chosen by Ogden for his sawmill, a spot chosen for "beauty as well as business." One description explains:

Smoke from a forest fire filters the sun's rays and bathes the area in a thick haze. Black Star.

. . . the river at this point runs through a slight bluff, which breaks into a low flat before the stream escapes from the borders of the town. . . . The business and residence streets were wide and well laid out, the houses prettily built and carefully painted, and little ornamental gardens were frequent.

William Ogden was greatly admired and respected, both in Peshtigo and Chicago. He was described as a "man of great public spirit, and in enterprise unsurpassed," and he had been Chicago's first mayor. His mind for business had made him a millionaire; his venture in the north woods had made him a lumber baron. Despite all his personal attraction, Ogden never married, preferring instead to live safely among friends in a "delightful residence" in Chicago's North Division.

Ogden built and owned the prosperous Peshtigo Company. The huge mill was steam-operated and contained ninety-seven saws whose blades ripped through 150,000 board feet of lumber each day. In addition, Ogden built a community around the mill for his employees. There was a company store and a boarding house for two hundred.

Many other buildings in the village were company-owned. The town managed to support two schools, possibly in the belief that there really was more to the world than chopping trees. Peshtigo had a population of two thousand, plus or minus a few hundred that drifted in during the peak lumbering season. Indians often took jobs at various mills in the area.

Ogden also built Peshtigo Harbor, six miles downstream. The harbor was always busy, with steamers taking on cargoes of lumber and sailing to Chicago.

The area had been slow to grow. When the Civil War came, growth came to a near stop as nearly all the men in the region left the forests and marched into battle. Wisconsin's population tripled, however, within twenty years, the lumber industry contributing greatly to its growth. Peshtigo had good potential for growth and was one of the three largest towns north of the city of Green Bay. Though dominated by the sawmill, the village of Peshtigo was coming into its own, with enterprising citizens starting their

own businesses. "As it stood, the pretty, bustling village combined the orderly enterprise of New England and the irrepressible vigor of the typical Western 'city'," boasted one report. A constant reminder of the future was the forest that surrounded the city, beginning directly at the village edge — "A solid wall of pine, oak, and tamarack," according to a contemporary description. The Peshtigo River divided the town in half.

Ogden, always with an eye for expansion, had encouraged the railroad to move north. Building had begun in early fall of 1871 and there were several railroad camps in the Peshtigo area by the end of September. Communications in the area were vastly improved by the addition of a telegraph line.

That fall the sawmills of Peshtigo continued their chore of buzzing through the pine, accumulating so many board feet of lumber a day, so many dollars of profit, and so many dollars to pay the hands that cut it. Waste of the trees was not the question — one-fourth of each tree was waste already. But there

was not time to develop improved cutting methods. The peak lumbering season — the harsh Wisconsin winter — was just ahead. There were between fifty and seventy-five lumber camps in the Peshtigo area during the winter.

The women of Peshtigo were beginning to can berries and other products for the winter ahead. Lumberjacks still spent their Saturday evenings and Sundays in Peshtigo at the various places of diversion provided there, for the winter lumber season would mean long days of hard work with no allowance for saloons and other pleasures. Though there was a strange foreboding circulating in the air that fall of 1871, the town continued to function as usual.

Peshtigo, located in Oconto County, was six miles south of Marinette, a small town located on the Wisconsin-Michigan border. Oconto County grew with the lumber industry and proudly claimed about fourteen thousand residents that fall. In addition to lumber, there was a substantial number of farms concentrated in the Sugar Bush settlements west of Peshtigo. The farms were for the most part isolated clearings. Should the woods around them hold any peril or natural disaster, the farmers and their families had to face it alone. The town of Oconto was about twice the size of Peshtigo and was located a few miles south down the lightly traveled stagecoach road. The city of Green Bay was located farther down that same road on the tip of the body of water called Green Bay. Wisconsin, a name originating from an Indian term meaning "gathering of the waters," reached statehood in 1848, and Green Bay was its second largest city in 1871.

The peninsula of land formed by Green Bay on the west and Lake Michigan on the east contained the most isolated farms in the state. Communities here were very small and few in number. The settlers, mostly immigrants who had heard that in America if you work hard you can be successful, came with a determination to make that success dream come true. Though their clearings and farms may have been poor and devoid of luxury, the small homes and barns were theirs, a possession they did not have before they came to Wisconsin. Door County made up most of the peninsula, with parts of Brown and Kewaunee counties on the south. The Belgians, Bohemians, Danes and French who had arrived in the 1850's and who had patiently and painstakingly built their little farms were at last beginning to profit prior to the fall of 1871. It had taken twenty years of hard work to attain the level of success they now enjoyed. The city of Green Bay was the principal center of trade for this area, but the road from the tip of the peninsula to Green Bay was a long one. Self-sufficiency, a phrase well-known by the farmers in isolated clearings, was the best answer to the isolation. It was, to them, survival.

The citizens in northern Wisconsin attempted to carry on business as usual at the beginning of fall 1871, amidst unusual conditions. The winter of 1870-71 had had less snow than usual and the spring thaw found streams and rivers without their full burden of melted snow to strengthen their currents. The less-than-normal snowfall was followed by a less-than-normal spring rainfall. By summer the area was gripped by a serious drought. There had been no appreciable rainfall since July 8, and swamps and creeks were dry. Indians still living in the area had never experienced such a lack of rain and felt sure that the cause of it all was the usurping white settlers.

By fall, the area was experiencing small fires that crackled and smoldered. The eyes of Peshtigo citizens were red from the ever-present smoke that choked the air. Some of the railroad crews found the situation serious enough to strike; though there was plenty of whiskey and beer in the saloons at Peshtigo, the railroad crews placed more importance on obtaining drinking water, a commodity becoming harder and harder to come by.

Fires ate at the humus underneath the bogs and in the forest floors throughout an area one hundred miles long and seventy miles wide. Occasionally the fires would surface and in early October of 1871 fires were emitting enough of a glow to be seen from the nearby small towns. Communities recognized that they were at the mercy of the forest and whatever mystery the resource chose to reveal. Preachers seized the opportunity that the events provided to warn that a land thriving too long with sin and without religion was to be dealt a serious blow. It was time to repent, they said. And though some probably did, the drought continued.

The telegraph lines had burned, leaving Peshtigo without a link to the outside world. Plank roads, constructed to expedite travel, had also burned. Smoke formed such a dense cloud over the area that captains on Green Bay, unable to gauge direction through their smarting eyes, navigated by compass though it was broad daylight.

Peshtigo was the only community north of Green Bay fortunate enough to have a fire engine. It was a hand-drawn apparatus proudly christened the "Black Hawk" and owned by the Peshtigo Company. Fire fighting there, for the most part, consisted of bucket brigades. Fire broke out in a few buildings that fall but everyone grabbed a pail and quickly filled it in the river, threw it on whatever happened to be burning, and returned to the river to repeat the procedure. When the burning swamps near the village became a problem, Peshtigo men determined that the situation was more serious and grabbed a shovel in addition to their trusty pails. Their efforts, coupled with favorable wind conditions, had managed to prevent any catastrophe. Part of the forest at the edge of the village had burned. This was viewed as a stroke of luck since it was believed that it would prove to be a natural barrier should the preachers' and the residents' apprehensions turn out to be true after all. As one person in Peshtigo said:

> ... The surrounding woods were interspersed with innumerable open glades of crisp brown herbage and dried furze, which had for weeks glowed with the autumn fires that infest these regions. Little heed was paid them, for the first rain would inevitably quench the flames. But the rain never came. ...

People of Peshtigo were still hoping for rain on Sunday evening, October 8. These were the people whose confidence had carved homes and prosperity out of a wilderness and they never gave up hope. Many were on their way home from church, "more promptly than usual" since the early October air was crisp. Some recalled that they felt occasional wisps of hot air in the smoke, to which they had by now become accustomed. The air was close and filled with ashes that swirled and eddied in the few puffs of hot air. A few citizens "delayed to speculate on a great noise which set in ominously from the West." The wind began to pick up and one resident remembered "the forest rocked and tossed tumultuously." The air became saturated with warmth — enough to nearly burn the skin. The noise became louder and completely impossible to ignore. Father Pernin, a north woods priest in Peshtigo at the time, later wrote that the deafening noise was like "the confused noise of a number of cars and locomotives approaching a railroad station, or the rumbling of thunder, with the difference that it never ceased, but deepened in intensity each moment."

Concern mounted as the people listened and looked at the familiarity around them, now hazy and yellowish. Some came from church, some poured from saloons, but all came to stand on the wooden sidewalks, to look at the towering, swaying trees and to ponder the immediate future.

Peshtigo was "declared by all" to be "the best built, most prosperous and happy town in all that

A fire storm from the heavens spread over the Peshtigo area with sudden swiftness, and isolated families in the thick north woods sought refuge in clearings. Painting by Milwaukee Journal *artist Mel Kishner from State Historical Society of Wisconsin.*

region." There was no happiness in Peshtigo the evening of October 8. The roaring noise increased and suddenly:

> ... a dire alarm fell upon the imprisoned village, for the swirling blasts came now from every side. In one awful instant, before expectation could give shape to the horror, a great flame shot up in the western heavens, and in countless fiery tongues struck downward into the village. . . .

In less than ten minutes after the first alarm, which occurred at nine o'clock, the entire village was engulfed in flames. A relief committee later recorded:

> ... The people came rushing in from the neighboring farms wild with fright, followed by cattle and horses in a confused rout. The

aroused inhabitants ran from their houses and their beds, and attempted to fly before the tornado of smoke and fire, which not only kindled into a flame the houses, but filled the air with flying bricks and timber. . . .

Some people fleeing in the turmoil and thunderous flames believed that this was truly the Judgment Day. Most fled with the thought that, Judgment Day or not, there was still a chance to outrun the flames, while others, just in case, repented as they snatched the hand of a loved one and fled. Still others preferred to calmly wait to be overtaken by the fire storm. A survivor later related to a reporter that it "was as like the Judgment Day as I can imagine. Friend Hansen, with his wife and four children, believed firmly that it was, and while the fire rained

down he began to walk composedly up and down his parlor with his family about him, and I have never seen him since."

Most people did not have Friend Hansen's composure. It was described later that the "tornado swept in currents and eddies of fire, in which many were caught and smothered on the spot, while others with great difficulty worked their way, some to the river and others to an open field on one side of the town." There was no time to grab pails and form bucket brigades; there was no time to ready the fire engine. Self-preservation was the motivation. Many intuitively believed that the large boarding house would not, simply on the basis of its size, succumb to the flames. Hundreds crowded inside but the structure was a mere matchbox beneath the roaring flames. After the disaster, the building was described as a "mass of ashes." None escaped from the boarding house and all that remained was "a pile of human ashes, from which can be picked out pieces of human bones, the largest not two inches long, and these split and broken."

Nearly all of the people fled in the direction of the Peshtigo River. It was first thought that the bridge led to safety, but Peshtigo was blanketed by flames on both sides of the river. People attempting to escape from either side collided on the bridge. Then the bridge caught fire at both ends and its human cargo flung itself into the waters below, and many drowned. Others gathered on the river banks and immersed themselves in the protective waters. There in the "red glare" they saw "the sloping bank covered with the bodies of those that fell by the way. Few living on the back streets succeeded in reaching the river, the hot breath of the fire cutting them down as they ran."

The timbers ready to be cut at the sawmill caught fire and floated among the people in the water. Many men, seeing the new danger, flung their coats "over the heads of wives and children, and dipped water with their hats. . . . Scores had every shred of hair burned off in the battle, and many lost their lives in protecting others." The onslaught of fire and wind was described by survivors as "instantaneous —

and the destruction almost simultaneous." Livestock swam in the river, crowding all who sought safety there.

"The flames played about and above all with an incessant, deafening roar," according to one report. In addition, "stifling currents of heat careered through the air for hours," claiming their toll in their super-heated air columns. A report taken from an observation of the scene the next day spoke of mothers and children, unable to flee the clutch of the fire, lying in "rigid groups, the clothes burned off and the poor flesh scarred to a crisp." People emerged from the cold water of the Peshtigo River at daylight, some having been in their watery refuge eight hours or more. Others, caught in the confused setting of fleeing livestock, human beings and burning logs, were unable to survive.

Peshtigo had been literally wiped off the map. Less than seven hundred people remained from the "happy and prosperous town" of two thousand. Survivors began the unpleasant search for relatives: Nearly everyone had lost someone dear. The dead were not always recognizable and lay in the streets where they had fallen. "Where houses stood," said an account, "the ground was whipped clean as a carpet and hope of identifying human ashes was idle." One horror-stricken relative found the remains of his nephew identifiable only "by a pen-knife embedded in an oblong mound of ashes."

On Monday evening, about twenty-four hours after the fire arrived at Peshtigo, the rain came, "gratefully to the living, and kindly to the fleeting ashes of the dead."

The night before was not devoid of heroism, though the fire struck with such suddenness and was of such proportions that people barely had time to save themselves. One farmer west of Peshtigo found himself in a clearing with his wife and fourteen children when the fire struck. The children were not all his — eight belonged to a neighbor who had sent them there for safety. The farmer kept his head as the flames rolled closer. With his hands (there wasn't time to procure a shovel) he dug and then covered all the children and his wife with dirt and then threw

handfuls on himself: All lived. A little girl who had managed to survive the Peshtigo fire was taken to Oconto and there well cared for by a generous family who adopted her. There was the story of a poor cobbler fleeing to the river who found the time to grasp the hand of a frightened child and bring her to the river. The child turned out to be a relative of the governor of Michigan who generously rewarded the cobbler's efforts.

Some preferred to die by their own hand rather than face the perils of a painful death by fire. One young man reportedly returned to his family's home, found his parents dead and slashed his own throat in despair. Another, apparently uncertain that his well was a safe hideaway from the sheet of flames, wrapped the bucket chain in a noose around his neck. One man, the search parties believed, had murdered his children and then himself after watching his wife die in the flames.

About 260 died in the Upper, Middle and Lower Sugar Bush settlements west of Peshtigo. Of these, many had been aware that certain danger would befall them. One Sugar Bush resident had declared that "all nature was so dry and miserable that it cried out for death." Farms were isolated and their occupants had nowhere to flee. The farms readily fed the flames. The desolation was so complete and the isolation so crucial that many who survived the fire starved to death.

Marinette had managed to successfully fight the fire, but only because it did not receive the fire's full force. The Marinette area, in fact, became a relief center after the catastrophe. At Marinette, the fire crossed the Menomonee River and raged through an area of Upper Michigan, adding more death and destruction. To the south, other cities had been hit by the flames but nowhere was the destruction as complete as at Peshtigo. Green Bay awoke Monday morning to find that the entire area to the north of the city had been destroyed and that nearly all of Door and Kewaunee counties on the peninsula had been flame-swept by another, separate fire. Green Bay was untouched and had complacently slept during the time the fires hit.

With one sweep of flames the fire on the peninsula had destroyed all the fulfillment of the industrious farmers. When the fire struck it was fought with small buckets of water. Its immensity was too much, unlike any other forest fire inhabitants of the region had witnessed. Buildings could not be saved; livestock was set free or left to burn in the barns. Often, the human occupants could not save themselves. Those who did survive lost all their material belongings. A later report of the destruction said that at least four hundred farms were completely stripped and desolate. It concluded that the losses "must foot up to one thousand dollars a family."

A committee from Boston came to organize help for the area and soon had on its books the names of "four thousand persons who are utterly destitute, who must receive aid until the next harvest." The proud independence of the farmers was recognized by the committee: "In order that the people may help themselves they must preserve their cattle."

Several hundred people perished on the peninsula. In one place "three or four children were found on their hands and knees, with their heads against a large stump, dead in this position." Most met their terrible fate without a struggle, probably killed outright by the first hot breath which they inhaled.

the boards that mark their graves are marked "2 unknown," "3 unknown," etc.

The injured crowded a nearby hospital and spoke of the tragedy. "Most of them suffer more from hurts of mind than body," it was reported. One woman kept hearing the screams of her crippled son and a "sprightly little girl" who perished in the flames.

Nearly 1,280,000 acres were burned over. To make things worse, communications were poor and it took some time for news of the fire to reach the rest of the country. A telegram from Green Bay did not reach the state capital of Madison until October 10. By that time, news of the conflagration at Chicago was known throughout the country. At the very time that Chicago was burning, several hundred miles to the north an even greater fire was taking place. It was America's greatest natural disaster, but, much to the resentment of north country citizens, it never got its due publicity and was always overshadowed by the Great Chicago Fire. William Ogden was victimized by both fires, losing a total of three million dollars.

At the time the telegram was received in Madison, Governor Fairchild and other high state officials were in Chicago organizing relief for that city. It was the governor's wife who, upon learning of the great Wisconsin fire, managed to divert supply trains northward. Many relief committees were formed and came to the aid of the Wisconsin fire victims.

People wondered why this fire had been so much more destructive than other forest fires. It was concluded that certain climatic conditions combined to form the right arena for the fire. It is believed today that the fire attained the dramatic proportion of a fire storm, much like the fire storms caused by incendiary bomb attacks in World War II over Germany and Japan.

Rebuilding was slow but deliberate. Today, the site of America's greatest natural disaster has about the same population as it did in 1871. A plaque at Peshtigo marks the graveyard of fire victims, and other trees have now replaced the pines. The once-busy Peshtigo Harbor is deserted except for the wildlife that inhabit the area, and the term "Peshtigo Fire" is thought of primarily in terms of its appeal to the tourist trade.

Williamsonville, a Door County settlement built around Williamson's mill, was wiped out, along with most of the Williamson family. Of the seventy-six people in the community, only seventeen lived, the great majority of them having perished beneath a blanket of flames in a fifteen-square-foot clearing. The nearest aid for the survivors was twelve miles northeast at Sturgeon Bay.

Between 1,200 and 1,500 people perished in the entire north woods fire on October 8. At Peshtigo it was lamented:

The names of half the dead will never be known. They are buried all over Peshtigo, and

Fire Equipment - Old and New

The first practical American hand pumper was designed by Thomas Lote in 1743 and copies Richard Newsham's London model. This 1855 hand pumper (opposite) employed the same principles. Nearly a dozen men moved the bars, or brakes, up and down to operate the pump. Steamers (above), drawn at first by four horses and later by three, replaced hand pumpers in the mid-nineteenth century. Even the first models could throw a three-hundred-foot stream. Both Jeff Kurtzeman and the Hall of Flame.

The first fireboat was built in 1809. Used in New York City, it had a crew of twenty-four to row it to the fire. Fireboats in the 1920's were capable of highly pressurized streams (opposite, top). James Haight. At fires away from the waterfront, it was motorized apparatus that literally passed the horse-drawn engines (opposite, bottom). Milwaukee Fire Department. Chemical carts also treated the water streams (above). Jeff Kurtzeman and the Hall of Flame.

Specialized equipment (opposite) provided the fuel for the modern fire apparatus. This old tank engine served between 1925 and 1949. James Haight. Pumpers, when coupled with a device that can support a stream at great height such as a water tower (below), are indispensable to the fire service. Some modern pumpers can pump over two thousand gallons per minute. Peter Pirsch and Sons.

Colorado Mining: Fortune and Fire

"There's gold in them thar hills" was a belief tenaciously clung to by cowboy Bob Womack as he rode the ranges of southwestern Colorado. Though at first known only for his wild demonstrations of horsemanship when he rode into town to celebrate on a Saturday night, Womack's casual prospecting and chance discovery of gold was celebrated nationally. For many years that area of the state had been a cattle-producing center, filled with herds placidly living off the abundance of the land. Cows were soon replaced by luck-seekers. By spring of 1891, because of Womack's discovery, the once lonely cattle range just southwest of Pikes Peak was filled with prospectors and speculators.

A mining district was organized in the fall and soon this was incorporated into Cripple Creek, a town named after a stream that trickled through the area. Bob Womack, after appropriately naming his claim Poverty Gulch, sold too soon for too little. His price was a mere $500, which gave him little more than a night on the town. He died in poverty in an area that produced twenty-eight millionaires. For instance, on July 4, 1891, Winfield S. Stratton, a carpenter from Colorado Springs, staked out a claim that turned him into a millionaire. Gold production passed the half million mark in 1892. More and more people flocked to the majestic, barren hills of Colorado and turned those hills into centers of excitement and activity. Cripple Creek became one of the richest gold camps the country had ever seen and soon a city was built. The lonely prairie that once belonged to the cows and cowboys now belonged to fortune hunters from all over the country. Gold output in 1895 was three times as much as it was in 1891.

Gold was mined and then spent. The people of Cripple Creek built fine businesses and created exciting diversions on which to spend their gold. They built grand hotels and banks. The city had newspapers and telegraph and telephone lines to tell the world of its great progress. Electricity lit the streets and homes and a trolley line curved through its streets. By 1894 Cripple Creek could boast eight hundred businesses and that did not count gambling houses, dance halls and other questionable places. Within a few months after the discovery of gold, lots jumped in price from $50 to $5,000. The area attracted sixteen doctors, thirty-six lawyers and forty-two real estate brokers in a total population of less than ten thousand.

For all its businesslike respectability, there was no mistaking this town for an ordinary town. It was a boom town and all conversation centered around gold. Gambling houses were open all day and on Sundays, too.

On April 25, 1896, the sun shone on the flimsy wooden buildings of Cripple Creek. Wealthy spenders filled the gambling dens and saloons. In a room over the Central Dance Hall a man and a woman were involved in an argument. Harsh words soon erupted into pushes and punches and in the turmoil, so the story goes, a kerosene lamp was overturned and immediately burst into flame. Seconds later, the fire was out of control and hot words were forgotten in hot flames. The man and woman ceased arguing and began spreading the message of "Fire!"

Somehow, in the haste of building a town, reaping its wealth and spending its money, Cripple Creek had

Cripple Creek, Colorado, could not suffice on one major fire in a week—it had to have two.
The west half of Cripple Creek burned three days after the east half was consumed in April 1896. Here,
the Denver House is being blown up to make a fire stop during the second fire. Denver Public Library.

remembered to provide for a volunteer fire department. But in a town with so many diversions — gambling, the arts, vice — the population was so busy diverting itself that it took some time for all the volunteers to drop their cards, leave the gambling wheel or abandon whatever other activity they were involved in. When the fire department finally arrived at the Central Dance Hall, its guests were already pleading for help from upper windows. Dance Hall girls casually slid down ropes or simply jumped. The roof quickly collapsed and the fire roared into the business district and danced up the hill toward the homes of Cripple Creek's millionaires. The nearby town of Victor sent fire apparatus but it was of little use. When the fire was out at last, thousands were homeless and nearly thirty acres of the town were

destroyed. Relief work was begun from instantly built structures.

But one fire was not enough for Cripple Creek, a town that pursued everything, even tragedy, with incomparable wholeheartedness. That same week the Portland Hotel caught fire from hot grease. The wood was dry and a prairie wind was whistling through the streets. The roof of the structure burst and showered Cripple Creek with little sparks. Wind toyed with the little fires. The boilers of the hotel exploded under the strain and so did a half ton of dynamite stored further down the block. This was enough for even the people of Cripple Creek. People mounted a large hill near a reservoir to wait for the end. When it was over, the entire business section had been destroyed. At the height of the fire, a stagecoach driver reported that he

saw flames one hundred feet high from a distance of ten miles.

All of Colorado aided in the relief work. Trains arrived and workers handed out medical goods. Those who were hungry and could not wait for the relief train plundered food stocks.

But the belief that "there's gold in the hills" still prevailed and Cripple Creekers were eager to get going again. One article explained that "the Cripple-Creeker was proud of his town before its destruction" and vowed to rebuild it into a town that would serve his prosperity even better. Fire limits were established. Buildings of material more flammable than brick or stone were expressly forbidden within fire limits. Only three or four buildings had withstood the fire. These had been made of brick and after the fire, brick businesses, homes and churches took the place of their wooden predecessors. New foundations were

begun at once, and within two months the finishing touches were being added. Improved architecture evolved into a pretentiousness that rivaled the Eastern cities. Cripple Creek now had several exquisite hotels and gilded saloons.

Labor troubles, however, were to become Cripple Creek's next serious problem. As more and more people came, the wages paid for mine labor began to go down. In 1894, mine owners announced the lengthening of the work day with no increase in pay, and there was a serious strike in 1903. Throughout Colorado there were continuing labor problems among miners and a tumultuous decade of violence. Fire and tragedy were destined to return to Colorado mining areas, although it would be coal rather than gold mines providing the backdrop.

The coal district of Colorado was primarily mined by the Colorado Fuel and Iron Company in which

The El Paso Livery Barn (opposite) exploded in flames during the second Cripple Creek fire. Denver Public Library. Other Western towns experienced destructive fires also. Victor, Colorado (above), had a devastating fire on August 21, 1899, which burned the Gold Coin Mine. State Historical Society of Colorado.

John D. Rockefeller owned controlling interest. In September 1913 between twelve and fourteen thousand coal miners went on strike. Their demands consisted of recognition by the company of the United Mine Workers (UMW) union, better hours and wages, and strict observance of state mining laws. The state militia was called out because both sides had yielded to the preparation of violence and were armed. It was reported that mine owners imported gunmen and that members of the union imported aliens to support the strike. Governor Ammons called many conferences on the situation, but nothing was decided and afterward there was violence — arson, murder and rapine. Mine owners argued that they were willing to make concessions but the union's "high-salaried agents and agitators, imported for the purpose" had turned them down. On the other hand, Frank Hayes, vice president of the UMW, insisted that state laws were being violated by the mine owners and that many miners had been killed for union

affiliation. The UMW accused the state militia of being a strike-breaking agency.

In early 1914 Congress launched an investigation in seven Colorado counties to determine if "the constitutional rights of citizens have been trampled upon." Troops in the areas investigated were on their best behavior. The committee left; nothing had changed. Federal troops were withdrawn from all the areas, except for one detachment that remained at Ludlow, less than 150 miles south of Cripple Creek, in the extreme southern part of the state.

When the strike began in September, the miners and their families were forced to leave company-owned housing. The UMW set up six tent colonies in the region, the largest one at Ludlow. The site was on private property leased by the union, which supported the canvas community of nine hundred people. The National Guard was called in to maintain order and, for added force, the mine owners hired professional strike breakers. The militia at Ludlow

Earthquake and Flames

was headed by Lieutenant Karl E. Linderfelt, described in a report by a law professor on a committee studying the strike as bloodthirsty and whose major purpose was "to provoke the strikers to bloodshed."

The tent colony was strategically located since it was near the Colorado Southern Railroad station where strike-breakers would detrain under the surveillance of the strikers. By April 1914, families had lived there nearly seven months since the strike began, trying to maintain a degree of normal living, but digging holes beneath their tents — just in case of serious trouble. Women and children made up two-thirds of the colony. On April 20 the National Guard demanded that a small boy be released from the tent city. Louis Tikas, a union organizer who was head of the colony, emphatically denied that anyone was being held unwillingly. Not long after the denial, dynamite exploded in the militia camp and the well-armed troops launched an attack. According to reports, militia machine guns were able to spray four hundred bullets a minute into the miners' tent homes. The fire was returned by miners, but their ammunition quickly dwindled. Plans were made by the strikers to run to the shelter of the railroad cars, while some of the women and children scurried into a ditch. Tikas was brutally murdered by militia bullets. One unarmed miner saw his son killed by a rifle bullet as he comforted his little sister.

The militia, seizing its advantage to do more destruction, set fire to the tent colony. Women and children crawled into holes to save their lives. The country intently watched the war in Colorado, waiting for accounts of the tragedy. The *Literary*

Rudyard Kipling, after a visit to San Francisco, wrote in 1891: "San Francisco is a mad city — inhabited for the most part by perfectly insane people whose women are of a remarkable beauty." San Francisco was a city that moved with activity, excitement and enterprise at such a fast pace that it might well appear to an outsider to be composed of lunatics nervously hopping in an unending quest to get things done. Not that the city lacked culture. But in opera, San Francisco taste demanded the vibrance and passion of *Carmen* rather than, for example, the spiritlessness of Goldmark's *The Queen of Sheba* which opened the season on April 16, 1906. In politics, San Francisco followed the shaky road of graft rather than the sure, but dull, path of honesty and responsibility. In materialism, it was "spend it while you can," for great fortunes had been quickly won and quickly lost and quickly amassed again in the city's short history. In commerce, the insecurity

of gold, silver and land speculation provided the seemingly necessary intrigue craved by the public. Immorality did not hide in the dark corners — a whole center for it, the Barbary Coast, was built with the help of City Hall.

The city, despite its location on a hilly site overlooking a beautiful bay, was not picturesque. Its many hills were lined with the shacks of immigrants and a large portion of the city contained the narrow streets and close buildings of Chinatown. Nob Hill supported the heavy ornate palaces of millionaires. And strung throughout the city was the confused network of cable cars introduced by Andrew S. Hallidie in 1873, greatly unappreciated by the people of San Francisco.

The Bay Area slept quietly unnoticed for years. It was not known to Europeans until the late eighteenth century and even then explorers sent out to mark the area sometimes unknowingly slipped right by it. But after California became the possession of the United States and gold was discovered there in January 1848, gold-seekers swelled the city for supplies and entry to land that beckoned with the promise of easy fortune. San Francisco extended its wharves into the deep water of the bay; marshy land was quickly filled and built upon. Many of those originally seeking gold stayed in the city to go into business serving the steady market of people pouring in. Chinese were imported to work on the transcontinental railroad which, when completed in 1869, brought additional commerce and wealth to the city. San Francisco was a dynamic, ever-changing city protected from gangs by vigilante committees in the 1850's and riddled by demagoguery in the 60's and 70's, with the ever-present corrupt city government working to its own advantage. By 1900, its streetcars were run by

The San Francisco earthquake rocked the city as it slept and buildings of poor construction, as these apartments, crumbled. Chicago Historical Society.

*The earthquake that prompted the San Francisco conflagration occurred at
5:13 a.m. on April 18, 1906. Overturned lamps and disrupted furnaces caused many
blazes and great clouds of smoke enveloped the city. Chicago Historical Society.*

electricity and in 1906 the population was nearly half a million.

The city experienced frequent earth tremors with more serious ones occurring in the years 1864, 1898 and 1900. Californians were used to tremors and were inclined to regard any slight shaking of the earth as "just another quake." At 5:13 on the morning of April 18, 1906, the people of San Francisco were awakened by a shaking and trembling of the earth that turned out to be more serious than anything the city had experienced before. Walls crumbled and chimneys toppled within. Whole tenements leaned to one side. Half-clad San Franciscans turned out into the streets, believing it was the end of the world. Though it had cost between six and seven million dollars to build, City Hall was so cheaply constructed that it crumbled nearly as quickly as the small wooden buildings that comprised ninety percent of

the city. The false land that had been filled to support buildings and trade was quickly filled again with the ruins of the structures. The quake lasted a little more than a minute. Many were trapped in their homes and many were buried beneath the wreckage, while others incredulously looked on.

Gas and electric mains broke, and were the causes of many fires which suddenly sprang up throughout the city. Lamps and stoves that tottered and fell during the quake caused other fires. The fire alarm system was crippled by the quake but firemen, many just returning from a three-alarmer, immediately responded. Fire Chief Dennis T. Sullivan was disabled by a chimney which fell on him during the quake as he slept, and he later died. But the entire 585-man department was soon on the various scenes of the fires under the direction of the deputy chief. At least thirty fires were in progress and that number soon

*This picture, taken by Arnold Genthe, called the "Father of Modern Photography,"
is rated one of the best news photos ever taken. Crowds line Sacramento Avenue on April 18
to watch the fire's progress. California Palace of the Legion of Honor.*

would nearly double. Fortunately, none of the department's equipment, including its thirty-eight steamers, was damaged in the catastrophe.

The city had an excellent fire department, but there were situations that even the heartiest fire department could not have handled. After the hose was coupled to the hydrants, doom was spelled out to the department and the city: The water mains had also been broken and there was no water to fight the fires. The National Board of Fire Underwriters had warned San Francisco of the disaster that would occur if certain standards of fire safety were not met. Chief Sullivan had the foresight to map a fire-fighting plan should a conflagration ever occur in San Francisco. It called for fighting the fire on one or two fronts; but several fires raged in the city and due to the lack of water the fire had to be fought on several fronts. The fire department drafted water from the bay or from sewers. Fires began to merge but still the department fought to contain them and to continue rescue efforts. By 10:00 a.m. three hundred bodies had been pulled from beneath the rubble.

The crowds were disorderly and panicky. Fire apparatus could proceed through the crowds only with great difficulty. Horses and automobiles were commandeered to get from one location to another. Mayor Eugene Schmitz, who was put in office by gangsters and who had, some contemporaries believed, plummeted the city's highest office to its lowest level of corruption, authorized that dynamite be used to make fire stops. He declared martial law and called in General Frederick Funston to maintain order. His Federal troops soon were patrolling the streets vowing to kill anyone caught looting "or committing any other crime." While many people sat on roofs or the hilltops watching the not-yet-threatening fire, others were fleeing ahead of the fire to Golden Gate Park, the Dunes or the ferries (the main link to the outside world) that carried them to the safety of outlying towns. The millionaires of Nob Hill, if their homes still stood, took in refugees. Soon food was being rationed under the supervision of Federal troops. People, true to the character of San Francisco, fled in a hurry with, as one correspondent

If Nero fiddled when Rome burned, girls laughed when San Francisco blazed. Arnold Genthe photographed this on Russian Hill. California Palace of the Legion of Honor.

The Hearst Building was a familiar landmark destroyed in the disaster. California Palace of the Legion of Honor.

described, "trunks, bird cages, sewing-machines, or whatever other treasures they valued most, then driven on from these places and trudging westward, like a retreating army, leaving their incumbrances scattered along the roadside, the whole population — cripples, invalids, children, and all — flowed toward Golden Gate Park and the Presidio."

The dynamite employed to eliminate fuel for the fire was unsuccessful. Buildings were blown up directly next to the fire and the rubble from the explosions merely provided more fuel.

There was no wind but convection currents were created sending flaming debris far ahead of the main fire. San Francisco's main thoroughfare, Market Street, and the entire business district were destroyed. At its height the blaze could be seen from fifty miles away. Fire fighters on the waterfront were aided by navy vessels, tugboats and a fireboat. They managed to stop the fire before it reached the rich wharf area. The *San Francisco Call* building was nearly twenty stories tall, one of the tallest buildings to be found in its day. But even its heavy construc-

It was difficult to ascertain how much of the damage was due to the earthquake and how much was due to the fire, as this view of Fulton and Leavenworth streets (above) shows. Since Market Street, the city's main artery, was 160 feet wide, the acting fire chief thought it would make a good fire stop. However, both sides of the street soon raged out of control (opposite). The tall San Francisco Call *building burned a floor at a time, from the top, down. Both Wide World.*

tion was no match for the flames. Glass melted and the fire, provided with the natural draft of elevator shafts and other openings, raged through the entire contents while the outside structure remained intact.

There were valiant efforts to save Government buildings. The U. S. Mint contained $200 million in gold coin. The building had to be saved. Employees and soldiers worked for seven hours under the command of a former Oakland Fire Department chief, and were successful. Meanwhile, post office employees beat the flames with wet mailbags. The Bohemian Club, a social organization whose membership consisted of the outstanding artists, writers and businessmen who had been lured by the romance of San Francisco, persistently fought to save the home

of Mrs. Robert Louis Stevenson, widow of the famous writer, and they succeeded. Many fine mansions were lost, however, including the Leland Stanford mansion on Russian Hill, which had been willed to Stanford University. Nearby a caretaker battled the fire for hours in an effort to save the estate in his master's absence. At last he decided to retreat and raised the American flag on the seemingly doomed property in a farewell salute. Army troops saw "Old Glory" wave in the distance and intuitively responded. The caretaker, with a whole army to aid him, saved the structure.

The poor Italian immigrants of Telegraph Hill fought bravely to save their shacks, using dirt, water and, as a last resort, their fine Italian wine. Though

the homes were mere shacks, in many cases it was all the impoverished newcomers had and they fought to the end to save them. There were many successes.

Though many were fighting hard to spare buildings from the fire, it was learned how quickly a simple act of carelessness could undermine all their efforts. Immediately after the quake a proclamation was issued ordering that no one light their stoves until the chimneys had been inspected for earthquake damage. Shortly before noon someone near the area of Market Street, either due to ignorance of the mandate or the natural desire to assuage one's appetite by cooking food, lit a kitchen stove. The chimney was indeed faulty, and soon the whole house was engulfed in flames. The fire department arrived but the fire was raging out of control and burning other houses in the neighborhood.

As the fires merged and grew more intense, firemen could only attack at outermost points. The fire continued its hold on the city until the department managed to make a successful stop at Van Ness Avenue on Thursday, April 19. Van Ness was one of the widest streets in the city and gateway to the Western Addition. Some buildings along Van Ness were quickly and coldly demolished with dynamite while others were sacrificed to fires insensitively, but scientifically, set by the troops and fire department. Nearly a mile of magnificent structures lining the street were given to ashes and rubble to save the homes of 150,000 in the Western Addition. A

favorable change of wind made the fire stop here. In propulsive San Francisco, the fires were not allowed to roam and passively burn themselves out, but were ambitiously fought on all fronts at all times until they were out. By Friday, all fires were under control.

San Francisco was a mass of rubble and smoldering ruins. Loss of life estimates went as high as five hundred and property damage was estimated at $350 million. Nearly seventy-five percent of the city, about three thousand acres, was destroyed. The infamous Barbary Coast and Chinatown, the country's largest Chinese community, were entirely gone. Almost 300,000 were homeless as the densely populated residential areas were destroyed. The scene was enough to prompt Lawrence W. Harris to write:

> From the Ferry to Van Ness, you're a god-
> forsaken mess
> But the damnedest finest ruins; nothing more
> and nothing less.

And they were fine ruins upon which a more earthquake- and fire-proof city would be built. Within a week business was being conducted upon the ashes. Aid poured in from America and Europe. For the next month everyone lived on free food, carefully rationed. Rigid sanitary rules were imposed. In addition people were not allowed to have lights in their homes nor allowed to open their own safes as they were excavated from the wreckage until authorities granted permission. Order prevailed. As one observer put it: "The old American rule that everybody can do as he will with his own was abolished, and people had to do as the sentry told them." On the other hand, only one insurance company folded; the others hid under "act of God" clauses or devised ingenious, but legal, ways to avoid paying what they owed on clients' policies.

Almost no traces of the disaster remained three years later. From those fine ruins, through the devotion and resolution of San Francisco's energetic people, emerged an even finer city.

Little remains of what once were comfortable homes (above) on San Francisco's hills in the Arnold Genthe photograph titled "Stairs That Lead to Nowhere." California Palace of the Legion of Honor. Ruins along California Street extend through Chinatown to the bay (opposite). The shell of St. Mary's Catholic Church, later rebuilt and still popular with San Franciscans, stands on the left at the intersection with Grant Street. Wide World.

Fire On the Waterfront

Four of the largest steamships of the North German Lloyd Steamship Company were docked at the line's newly remodeled and enlarged piers on a pleasant Saturday, June 30, 1900. The docks were located on the north Jersey shore of the Hudson River at Hoboken. New York City was on the opposite shore. There was nothing unusual about the construction of the docks: Pilings had been driven over the entire area to be covered by the piers and crowned with large twelve-by-twelve timbers. The joists and wooden flooring rested atop the timbers. The sun was shining and the ferries on the river were crowded with passengers either taking a leisurely ride or escaping from the city to the country. People in the pursuit of relaxation or a simple breeze crowded the shores.

There were a few on that Saturday who elected to visit the North German Lloyd Company's ships to get a first-hand glimpse of the huge, floating freighters. The *Kaiser Wilhelm der Grosse*, which was the largest and attracted much attention, was moored at Pier No. 2, loaded and prepared to leave the following Tuesday. The *Kaiser*, though a freighter, was also a popular passenger vessel, and one of the fastest, holding the speed record across the Atlantic: twenty-two-and-one-half knots an hour. The S. S. *Main* was docked at the new pier, loading to return to Germany. Next to the *Main*, at Pier No. 1, was the steamer *Bremen*, a German-made ship with a register of ten thousand tons and a speed of fifteen knots. Pier No. 2 also accommodated the *Saale*, an older steamer. Just a few hours before, the *Aller* had quietly sailed from the docks. Pier No. 3 was host to several lighters (boats used to carry freight throughout the harbor) and barges. All of the steamers at the docks were taking on coal and cargo.

A fire began in the large stacks of cotton bales piled on Pier No. 3 at about 3:55 p.m. At the first sight of smoke an alert person on the docks had turned in an alarm, but it was already too late. Flames leaped to heights of one hundred feet. The fire, with the help of the wind, spread rapidly and discovered an explosive amount of fuel in the whiskey casks stored on the dock near the cotton. With a violent roar, the whiskey casks burst, and about two hundred longshoremen on the pier made a desperate dash to escape. The fire followed hot on their heels. About forty men were not quick enough and perished on the pier; others hid under the pier, clinging to the supports, only to be suffocated by the choking flames.

The many passengers and visitors on board the ships were swept by a wind of fire which cut off their escape. They were unaccustomed to the discipline and procedure on ships and struggled in panic through the pandemonium. Some people on the upper decks managed to scramble down gangplanks, but the crews and passengers on lower decks of the vessels could not escape, since the fire was consuming their escape routes. Men on upper decks were grasped with torments of confusion. Some fought and clung to each other, even though to do so meant that no one could jump and all were doomed to death by fire.

Cause of the fire was speculated to be a careless match or cigarette, or a liquor barrel explosion. The cotton on the pier could have been smoldering unnoticed for days and when conditions were right, come to vivid life. Whatever the cause, eyewitnesses stated that within nine minutes from the time the blaze was first seen, the piers and four steamships adjoining them were burning fiercely. One report said that in less than fifteen minutes flames covered one quarter mile.

The fire was huge, and as the curtain of smoke went up, half a million people crowded New York rooftops for an afternoon of spectacular theater.

Seattle is a major port and does much shipping to Alaska and the Far East. Here, a fireboat plays a steady stream in the attempt to douse the fire on the Grand Trunk Docks, July 30, 1914. University of Washington.

Tugs work to free the S. S. Bremen (left) and the S. S. Main (right) from their berths at the Hoboken docks in an attempt to save them by sending them to sea. While fireboats toiled alongside, fire fighters' efforts saved one hundred people on the Bremen. Later, both beached upstream at Weehawken.

Some jammed ferry boats or congregated on the shore. A reporter compared the size of the crowd to the throngs that had gathered to welcome the hero of Manila Bay, Admiral Dewey, back to New York in 1899. It soon was obvious, though, that this was no gala reception. Confidence was placed in the heroes of the drama — the fire fighters, but this was a tragedy, and the heroes' tragic flaw centered around not being able to get to the fire in time and then not being able to get close to it. Nearly 1,400 people were the innocent protagonists threatened by the villainous fire, one of the world's most ancient and most experienced antagonists. Firemen were further crippled by a lack and loss of hose. The Hoboken Fire Department was unable to gain on the flames. Adjoining piers were dynamited to withdraw potential fuel for the fire. Firemen from Hook and Ladder Company No. 2 of Jersey City, a few miles to the south of Hoboken, narrowly escaped death while rescuing people from beneath the pier just south of the blazing docks. Flames suddenly flashed in front of them and approached them, cutting off their escape. Fortunately, the wind shifted and drove the flames away.

Hundreds sought escape by jumping into the river, but the river itself appeared to burn. One witness said, "The surface of the water was covered with floating and blazing masses of freight thrown in haste from the doomed vessels." Tugboats steamed to the area, some to stay close to the doomed ships to enrich themselves with spoils of the ships' cargoes. Most of the tugboats in the area, however, concentrated on saving the people who had jumped to the safety of the river. Fire Chief Applegate undauntedly pursued rescue operations on the pier, in the ships and in the water. One fireboat attempting to get a stream on the *Bremen* was forced first to threaten to turn its water on the tugboats that vulturously hovered near the ship's hold. Despite the greed of some, however, one hundred people were saved from drowning. Saloons and stores on the waterfront were turned into temporary hospitals.

Most of the property-saving efforts were concentrated on the *Kaiser*. The mighty ship was towed onto the river with all decks ablaze. Efforts to save it were highly successful: Of 450 on board, not a life was lost and property damage was relatively light. The *Kaiser* sailed on schedule the following Tuesday. The other great ships did not fare so well. The *Main, Bremen* and *Saale* were flame-swept from end to end. Tugboats towed the *Bremen* and *Saale* onto the river, but the *Main* could not be freed from the pier. Passengers on board the *Main* fled to the burning decks and piers or were forced to jump overboard. All four ships were the finest that money could build and all burned in the presence of the most scientific fire-fighting force in the world.

The imploring faces of those on the lower decks appeared at portholes, which were only wide enough for a person's head to fit through. These faces eventually turned a lifeless gray, and many people had their heads and their arms hanging through the portholes. Some that had been waving white handker-

chiefs hung limp as the fire took its toll. The many boats in the area were helpless. Rescuers tugged at the trapped victims in the hope that the steel sides of the boat would somehow burst, freeing them. But it was no use; human bodies could not fit through the tiny holes.

On the *Saale*, trapped people begged for help at the portholes for three hours. At one point, firemen managed to contain the fire in one section of the craft and forty stokers were saved. It was hoped that many lives could be saved as the fire diminished and the ship grounded in the mud off Ellis Island. Human hands can tame fire, but they cannot stop the tide. The ship was tilted and the portholes were near the water. Those who had not perished in the lower holds due to fire soon felt the lapping of the waves and the encroachment of slow rising water overtake them. The majority that died on the *Saale* were coalmen and firemen. Just before the *Saale* filled with water a

woman had tried, though she was hot and tired, to calm the people around her. She tried to reassure them that though they had been waiting for hours, they would be rescued. But the small portholes proved to be too much for rescuers. The crew of the fireboat *Van Wyck* could do little exept to pass cups of water through the portholes to the inmates who cried for a drink of water "for God's sake." A tugboat pulled alongside the mortally stricken ship, and a priest on the tug gave the *Saale*'s passengers absolution.

The two most powerful fireboats in the world, the *Van Wyck* and the *New Yorker*, continued to pour a dozen pressurized streams onto the North German Lloyd ships. Tugboats still continued to work at freeing the *Main* from the dock. When at last the boat was freed and prepared to be led to the river, cries of trapped people from below, still alive, filled the smoky air. However, before the ship's skeleton was

The burning S. S. Saale (above) is towed from the Hoboken docks in an effort to save her. Between three and four hundred people died in this 1900 fire, but forty-seven years later an even greater tragedy occurred at Texas City, Texas. The Monsanto Chemical Works (opposite, top) caught fire after an explosion in a docked ship's hold. Wide World. Injured victims of the fire are carried out of danger (opposite, bottom). Others were not as lucky: Over one thousand people perished. Texas City Daily Sun.

led away from the dock, sixteen coal passers had been rescued.

The *Bremen* and the *Main* beached side by side off Weehawken, a town a few miles upstream on the Hudson River. One tugboat managed to save 104 men from the burning *Bremen*. Other tugs concentrated on preventing the New York docks from burning, and attempted to turn burning lighters away as they drifted into the New York harbor.

The fire advanced rapidly on the ships because there was no steam in the boilers to fight the fire, except that which was needed to operate hoisting engines. Many victims drowned in the river, and many others trapped on the ships could have been rescued if the portholes had been larger. Property damage was heavy with the steamship lines losing between five and ten million dollars. Other piers belonging to other lines had been lost as the fire spread in its destructive path. It devoured stores and warehouses which had cost over a million and a half dollars to build.

At 2:30 a.m. the next morning the fire still burned. As dawn arrived, bodies were being removed and it became clear how triumphant the fire had been. Most of the victims of the tragedy were crew members, many from other countries. The *Saale*, which had been ready to sail in the afternoon to Boston with 450 on board, had bodies strewn about the deck. The *Bremen* carried three hundred men, while the *Main* had 250. Few had escaped. When the fire finally subsided, between three and four hundred people were estimated to have died. Captain Mirow of the S. S. *Saale* died with his ship. The burning of the North German Lloyd docks at Hoboken, New Jersey, was a tragedy that had no equal on American docks until April 16, 1947.

Shortly before that day, the S. S. *Grandcamp* docked at Texas City, Texas, near Galveston, with a cargo that included peanuts, cotton, oil well machinery and sisal twine, to take on board a heavy

(continued on page 99)

Today the familiar steam fireboats are being replaced by the more economical diesel- or gasoline-powered crafts. A small motorboat containing deck nozzles and thin hose is often used on pier fires. Chet Born, San Francisco Fire Department.

The French ship Normandie, *once the world's largest, was a luxury liner embellished with the finest of details. The U. S. Navy purchased the ship in 1941, intending to convert it into an auxiliary transport. On February 9, 1942, while crews were at work on the vessel tied up at a New York pier, she caught fire. Smoke from the blaze blackened the Manhattan skyline. Wide World.*

cargo of fertilizer. Fertilizer brings to mind pastoral visions of tiny green plants growing strong and healthy. However, with the introduction of ammonium nitrate, which the *Grandcamp* was to load, fertilizer brings to mind a new vision: explosion. By nightfall on the evening of April 15, 1947, 1,400 tons of fertilizer had been packed into the ship's lower decks. The loading continued the next day and, while this was going on, it is believed that someone lit a cigarette. A few minutes later trickles of red-orange smoke came from the direction of the hold. Shifting of the bags of fertilizer revealed a small fire there. An officer ordered everyone off the ship and ordered that no more water be used for fear that the cargo would be damaged. The fire department was called to the heavy industrial area near the pier.

The ship exploded at 9:12 a.m. with a tremendous roar and killed four hundred people outright, including most of the fire fighters. Airplanes literally were blown from the sky, and a tidal wave was created which lifted a docked barge several feet inland. Children located over a mile from the blast were injured. The Monsanto Chemical Works near the docks caught fire; oil storage tanks blew up. Fire fighters, who could only approach the area wearing gas masks, attempted to free the ships and send them out to sea.

The blaze became uncontrollable and firemen were forced to withdraw. Fire raged throughout Texas City for two days. Death tolls could only be guessed since many of the victims were migrant workers, but one estimate was that one thousand people died.

SOS at Sea

"Such a disaster could hardly happen at sea," said a writer describing the tragic dock fire at Hoboken, New Jersey, in 1900. Yet on June 15, 1904, over a thousand lives, more than half of them children, were lost by fire on the New York excursion boat, the *General Slocum*. Since it was a work day, most of the passengers were women and children. The trip was an annual Sunday school outing and all had been looking forward to the ride on the large, wooden paddle steamer. A band even played as the ship set out down the river. As the boat traveled, people on the shore noticed that the ship was on fire. The passengers, however, did not detect it for some time.

The large loss of life was partly due to the complete lack of good judgment on the part of the captain. He headed the boat straight into the wind, fanning the fire. There was apparatus waiting on the shore to help fight the fire, yet the *General Slocum* continued to steam away. Many of the women and children who previously had been having a good time, listening to the music and catching the summer breeze, now were jumping overboard or being burned by the fire. Many were later found caught in the great paddle-wheel.

Attempts had been made to put out the fire. The crew members worked hard to extinguish it, but the equipment on board was useless. Poor leadership, rotten fire hose and lack of proper life preservers took its terrible toll.

The S. S. *Morro Castle* was quite a different story. The ship had elaborate fire-detecting systems. It was a huge ship, weighing 11,500 tons, a luxury vessel sailing between New York and Havana for the benefit of the fortunate few who could afford its pretentiousness at a time when the rest of the country was in a severe depression.

On the stormy night of September 8, 1934, the *Morro Castle* was steaming toward New York on its return voyage from Cuba. The evening before had been saddened by the death of the ship's master, Captain Wilmott, who collapsed from a heart attack. The acting captain was William F. Warms.

Smoke envelops the Morro Castle *as the fire spreads through the ship.* New York Daily News.

Fire swept the luxury liner quickly and without warning. When it was over (above), only twisted, writhing metal remained of what once were elegant staterooms. New York Daily News. The traditional law of the sea that women, children and passengers must be saved first was selfishly tossed aside: Of the first ninety-eight to be rescued from lifeboats (opposite), ninety-two were crew members. Wide World.

A fire, reported at 2:56 a.m., allegedly had begun in a deserted writing-room locker. The room was equipped with a fire door which was, for some reason, opened. Had the door remained closed the chances of containing the fire in one room would have been very good.

George Alagna, the second radio operator, waited at his post for orders from the captain to send out an SOS. None came, so Alagna went to the bridge to ask for his orders. He reportedly found Captain Warms in a befuddled daze. The radio operator was unable to get any orders from his captain and returned to the radio room to report to his immediate superior, Chief Operator George Rogers. Alagna's report described the people on the bridge as "madmen." He claimed that he made no less than five trips to the bridge before he managed to get an order for the sending of an SOS. This involved a risk of his own life. It also involved a crucial delay of twenty to thirty minutes before the ship's fate and position were made known to the other ships in the area as well as to authorities on shore.

Though the *Morro Castle* was equipped with the most modern fire-detecting devices, the 240 crew members on board exhibited a profound lack of knowledge of their proper use. Simultaneously with the sounding of a general alarm, the policy of "every man for himself" was adopted. Discipline among the second-rate crew was lax and any sort of preparedness for an outbreak of fire was obviously missing. There was no efficiency and seemed to be no authority. It was reported that many on the lowly paid crew were dope smugglers. Courage was a quality that had never been expected of them and it certainly did not reveal itself during the fire. Officers and hands alike thought only of their own safety, completely disregarding the *Morro Castle*'s 318 passengers. Crew members who hours before had been the slave of the affluent, now discovered that fire equalized rich and poor and all viciously clamored for their freedom. Men fought and trampled each other in the fight to get up a narrow stairway that led to safety.

The Ward Line that operated the ship was greatly concerned with its profits, and profit-making was achieved at the expense of the crew and, ultimately, the passengers. Conditions were poor on all the ships in the American merchant marine, not just the *Morro Castle*. The crews were overworked. They put in long hours for low wages and the drudgery was continuous. A ship sailed into port, then immediately set out again. Demoralized crews like the one aboard the *Morro Castle* were almost inevitable. The Ward Line specifically pursued an adamant anti-labor policy.

So there was no serious attempt to put out the fire on the *Morro Castle*. People fought to board the lifeboats, but as flames overtook the ship, many jumped into the waters and drowned. Many were incapacitated by consumption of too much alcohol. One report of the tragedy mentioned that the New York-Havana cruise was noted for its drunkenness.

The Morro Castle *is escorted out of the New York harbor on her maiden voyage to Cuba on August 21, 1930. The proud liner set a new speed record on that journey.* New York Daily News.

In all, 137 lives were lost. A shocked public demanded to know why. A special board for the Bureau of Navigation and Steamboat Inspection investigated the tragedy. The "human factor" received much treatment in the inquiry since it was known that the ship had up-to-date apparatus. The undependable human element failed in time of need. The crew and authorities saved only themselves. Testimony indicated that efforts were made to extinguish the fire but officers were not on hand to direct the operation and the efforts were quickly abandoned.

The Ward Line, in an attempt to free itself from blame (and also from the indemnity it would be forced to pay if found negligent), attempted to blame the fire on arson. What was more, these purported arsonists were *communist* arsonists, according to the company. It was known that there was labor trouble on board, so the company tried to breach the

credibility of the radio operator's testimony because Alagna had been active in organizing strikes for better working conditions.

A big question in the investigation concerned the delay of the SOS. No call for help was sent until many passengers had already jumped into the sea. Despite the stormy and rainy weather the fire could be seen seven miles away on the Jersey shore. Boats far away inquired about the fire long before a call for help was sent. No one can know what the captain was thinking. He knew that the Ward Line had to pay salvage fees on each SOS call that went out, so perhaps this was a factor.

It was learned that Chief Engineer Eban S. Abbott escaped in Lifeboat No. 1. His conduct could be classed as unheroic at best. He never went below once the fire was discovered but instead turned his responsibilities over to others ordering them to stay with the fire, while he quickly escaped all danger. Of

After the fire on September 8, 1934, that killed 137, the charred corpse of the doomed ship was beached on the New Jersey shore. The senseless loss of life could have been lessened had these three lifeboats been used. New York Daily News.

the first ninety-eight to escape, ninety-two were crew members. Lifeboat No. 1 carried thirty-one crew members and one passenger. Another carried nineteen crew members and one passenger.

All familiar courtesies and acts of courage and laws of the sea were simply turned aside. Someone reporting on fires at sea in a British magazine in 1852 saw gallant unselfishness prevail at the outbreak of fire and commented:

> ... and what a contradiction to the selfishness of humanity, to behold the first impulse of those in command in such circumstances directed to the preservation of all under their charge, refusing to avail themselves of the means of escape within their reach, till every individual has been cared for.

Such a contrast to what occurred on the *Morro Castle*! The captain's inexperience certainly played an important role. Passengers were not even helped into boats. There was a complete disregard of the ancient law of the sea that says that an officer should take his post and command his men in an emergency.

As a result of the investigation, Captain William F. Warms and four other officers were charged with negligence. Captain Warms was charged with delaying the SOS call, failure to stop the ship even after it was obviously on fire, and failure to direct effective fighting of the fire. Charges brought against the chief engineer, assistant engineer and second and third officers had to do with the fact that they saved their own lives without regard to saving the lives of the passengers.

Captain Warms was sentenced to two years in jail and Chief Engineer Abbott was sentenced to four years. But the sentences were overturned by a higher court and were never served.

The Triangle Shirtwaist Tragedy

Times were changing. It was the turn of the nineteenth century and women were becoming aware of themselves. Many were becoming involved in the political workings of the country; many were entering the work world. Women's fashions changed with their new freedom, and shirtwaists were very popular. A shirtwaist was a tailored, fitted shirt usually worn with a simple skirt. These crisp, new garments could be found in businesses throughout the country, worn by women seeking fulfillment and independence.

Factories produced the shirtwaists by the thousands. But the women who cut them, sewed them and produced them were not the poeticized new females. At the Triangle Shirtwaist Company in New York City, for example, women were underpaid and worked in unsanitary, unsafe conditions. Most of them were Jewish and Italian immigrants. Often they were unfamiliar with the English language, and were considered by their employers to be of inferior intelligence. Often these working women, many of them young, provided the only means of support for their families, earning about six dollars a week. At Triangle, members of the same family in many cases were part of the company's work force. They worked in quiet drudgery six days a week and were expected to do so with unswerving docility.

In late 1909 garment workers at the Triangle Shirtwaist Company organized a strike and there was violence; pickets were beaten and arrested. Many workers crossed picket lines and ultimately the strike failed. Afterward, women worked without complaint if they wanted to survive.

The factory occupied the eighth, ninth and tenth floors of the Asch Building on the corner of Washington Place and Greene Street. Triangle was one of many loft factories in New York City. The factory was only a block away from Washington Square Park where, on Saturday, March 25, 1911, children were playing in the new spring air. The scene was peaceful.

The Asch Building was completed in 1901 at a cost of about $400,000. Installation of a sprinkler system would have cost an additional $5,000, so it was omitted. The building was defended by its owner, Joseph J. Asch, as being fireproof. Building codes allowed the structure to have wooden floors and wooden window sashes, rather than metal. The Asch Building was supposed to have, according to the building code, three staircases. The building architect argued that though there were only two staircases, the fire escape could serve as the third. The Greene Street staircase was the only one with an exit to the roof — the Washington Place stairs went up only to the tenth floor. The single fire escape in the rear of the building reached down only to the second floor; below this terminal point was an enclosed courtyard. All the doors in Joseph J. Asch's building opened inward. In 1909, a revision was proposed in the building codes. This revision would have required a fire escape on the front of the building. However, the fight was lost in a political battle of special interests and was not passed.

Saturday was pay day for Triangle employees. On March 25, 1911, at quitting time, 4:45 p.m., a few employees on the eighth floor pushed back their chairs and walked toward the Greene Street freight elevators and stairs. Before reaching the elevators the workers had to pass one at a time through a narrow passageway. As they passed, they opened their purses so that the watchman could inspect the contents. Theft was reportedly a serious problem and the management wanted to insure that no shirtwaists or other company property went home with the workers. Isaac Harris, one of the owners of the Triangle Shirtwaist Company, dealt seriously with pilferers. He saw to it that they were arrested or fined, but some of the victims of his accusations had sued him. After

(continued on page 109)

The fire department arrived at the Triangle Shirtwaist Factory as the workers began to jump. Firemen first had to remove the bodies from their hose lines before they could tackle the fire, but the tragic blaze was quickly controlled. The building had long been recognized as a firetrap. Culver Pictures.

America depends on firemen whenever life and property are threatened by fire. Despite quick response, 145 people died in the Triangle Shirtwaist fire. Industrial fires often require multiple alarms to control them. A factory that is enveloped by intense flames and often accompanied by loss of life is quickly surrounded by firemen who attempt to put the blaze out. James Haight.

that, he merely dismissed them. How much was actually taken? Harris himself revealed later, after the Triangle tragedy, that the total value of the stolen goods amounted to less than $25.

Windows on the eighth floor opened in the back to the single fire escape, and there were also windows on the Greene Street side and the Washington Street side. Throughout the floor were long cutting tables. Beneath them were rag bins to collect scraps from cutting, and above were wires that were strung to hold patterns once they were cut. Workers were casually on their way to the exit when someone smelled smoke. One woman turned to look back at the cutting tables and saw red-orange flickers through the openings beneath the tables for the rags.

There were cries of "Fire!" throughout the shop and alert cutters were already emptying buckets of water on the flames. But the materials the flames were feeding on were more flammable than paper. It was already too late for buckets of water to do any good — the flames had discovered more fuel in the hanging patterns. Workers jammed the narrow doorway. The line containing the flaming patterns burned and fell onto the machinery. The fire was spreading with a rapidity that could not be contained; the flames reached the ceiling and the windows popped from the pressure within.

Someone located a fire hose and tried to put it to the flames, but no water came. The hoses had been allowed to rot and the valves that controlled the water supply had rusted. Flames were consuming the workers' tables. Screaming and frantic, the terrified employees rushed to the exits and crushed against the elevator doors. But the elevators could scarcely accommodate more than a dozen people at a time. Some jumped into the elevator shaft and, as the car went down, some jumped onto it. A few somehow managed to live.

One person on the eighth floor tried to warn the ninth and tenth floors. She was able to telephone the tenth floor, but due to a communications failure, she was unable to reach the ninth floor. Business went on as usual on the ninth floor — fire was the furthest thing from their minds.

Others on the eighth floor ran to the Washington Place staircase. Many tumbled down the stairs with the crowd, unable to remember later how their lives had been saved or just how they had made it. A policeman arrived on the scene and ran up the Washington Place stairs. He passed the confused, panic-stricken women, recklessly trying to make their way down the stairs. On the eighth floor he could see the flames attacking the women trying to get to the

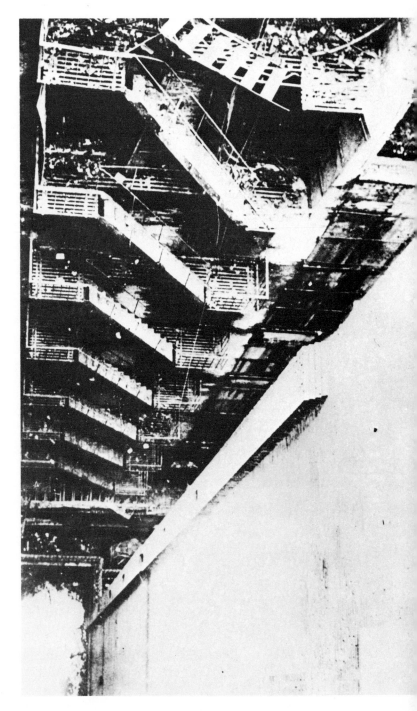

The Triangle Shirtwaist Factory's fire escape (above) was shaped into contorted steel by the heat, hurling its human load into the courtyard. After the fire, burned sewing machines (below) rested among charred ruins. Both Wide World.

*A tank truck loaded with twelve tons of gasoline overturned and
exploded in flames against the side of a Milwaukee food processing plant in 1967.
Unusual fire tornadoes or "funnels of fire" broke out during the blaze and posed
serious threats to firemen's lives. It took twelve hours to bring the fire under
control and a few days more to put it completely out. James Haight.*

staircase, and, with the help of an eighth floor machinist, he hustled them down.

A few on the eighth floor had located the fire escape. These stairs were narrow and perilous, and dangerously close to the burning building. Still they attempted to make their way down and save their lives. Some climbed into a broken sixth-floor window. Once inside, they piled behind a locked door on the Washington Place stairwell. The patrolman who had organized the exodus from the eighth floor heard the screams on the other side of the door in the sixth-floor landing. He was able to unbolt the door but he could open it only with difficulty because the door opened inside, toward the women, and they were tightly jammed against it. At last the door opened and the employees had the chance to fight their way down the inside staircase to freedom.

The people on the tenth floor rushed to elevators. Max Blanck, Triangle's other owner, was on the tenth floor with two of his children. He and the other workers on the tenth floor determined that the best and safest way out was to use the Greene Street stairs to the roof. The women wrapped coats and muffs around them and ascended the staircase. Flames shot out at them and many, when they reached the roof, had burned hair and clothing. There were about seventy people working on the tenth floor and all were saved except one.

On the ninth floor, smoke could be seen rising from below, and the first realization of fire came over the approximately 250 employees working there. They began pushing to the Greene Street stairwell exit. Here, too, they could pass only one at a time. Fire has the inherent ability to cause panic and already at the rear of the crowd people were pushing to get out.

The rickety fire escape was covered with frenzied workers trying to flee. As they descended, flames nipped at them through the windows, setting their hair and clothing on fire. Beneath the fire escape's end on the second floor was a glass skylight of a ground-floor extension that jutted from the Asch Building into the courtyard. The intense heat from the fire caused the metal of the fire escape to bend and it loosened from the building's heated walls. People smashed through the skylight as they left the

shaky security of the contorted stairway. The court-yard became a death pit.

The windows of the Triangle factory's three floors were shuttered with sheet iron. As the women desperately tried to free themselves by means of the fire escape, the shuttered windows bent outward in the intense heat, blocking the already disfigured stairway. The fire escape was increasingly clogged and burdened with its load. Some of the workers turned back to go upward, but flames hunted them from the tenth-floor windows. Passage that way was impossible, and a shutter jammed the eighth-floor landing. The fire was most intense between the eighth and tenth floors, so that is where the panicked, screaming women were trapped on the fire escape waiting for the inevitable. The flames leaped out from all the windows. Finally the fire escape buckled and swung, flinging its flaming human load into the courtyard.

The means of escape from the ninth floor were rapidly diminishing. Fire blocked the entrance to the Greene Street stairs so they could no longer be reached. Some employees ran to the Washington Place stair door and beat on it. It was locked, and their screams would not unlock it. Someone smashed a window, but this only allowed more flames to rush in. Women were battling for their lives without any idea of what they should do. Despite encouragement from the New York Fire Department, fire drills had never been held at the Triangle Shirtwaist factory.

A solid wall of flame now occupied the ninth floor. Workers were trapped in the long narrow aisles between the machines. Wicker work baskets served a dual purpose: They fed the flames, they trapped the workers. Fire was pushing the workers back toward the Washington Place windows. Flames lunged at them — there was no place they could go to escape. So they began to jump from the building, nine floors above the ground.

The first alarm had been turned in to the fire department at about the time the people began to jump. Life nets were stretched, but they could not hold human beings that had jumped from a narrow ledge nine floors up. And the people could not wait for the nets to be stretched. They were burning and the flames drove them into midair. At first they looked like bundles of cloth dropping from the windows. The crowds on the street below were shocked when they recognized the unfurling bundles as people. A ladder was raised, the tallest ladder in the New York Fire Department, but it reached only to the sixth floor.

Women jumped with their hair in flames. Some who had wrapped cloth around themselves for protection now found it burning around their bodies. The workers did not jump individually — girlfriends held hands and jumped together, men and women embraced, and, at one point, five people held hands and jumped. Tarpaulins were located and stretched, and the hands of the men that held them bled with the impact of so many bodies. Bodies crashed through the glass deadlights (shutterlike coverings) over the vault beneath the sidewalk and piled up. They were hitting all over, and the firemen had to clear the bodies away from the hose before they could be used. Workers finally arriving on the ground floor after descending the Washington Place stairs were not allowed by the firemen to get out, for fear that they would be killed by falling bodies.

It took only eighteen minutes for the fire to be brought under control. But when it was all over, 145 were dead. The impact of the tragedy was felt around the world, for many of the victims still had strong ties with their families in other countries.

Many questions had to be answered: Why was the building not properly equipped for an emergency? Why had a reasonably wide exit been partially blocked to allow insensitive employers to scrutinize for practically nonexistent theft? Why had orders to improve the building's safety been ignored? Why hadn't building inspectors and other responsible people forced compliance? And why had the Washington Place stair door been locked on the ninth floor?

After the initial accusations, charges and counter-charges, Triangle's two owners, Max Blanck and Isaac Harris, were tried on charges of first- and second-degree manslaughter. They were found not guilty. Many of the jurors had no doubt that the ninth-floor Washington Place exit was locked, but they could not prove that the girls did not know the door was locked.

Yet the tragedy was not forgotten. Sweatshops were reformed, and the International Ladies' Garment Workers' Union became strong. And the commission that rose from the Triangle Shirtwaist factory ruins did all it could to insure that those who had died had not died in vain.

Their hose snaking about the grounds, firemen decide on strategy to subdue the flames. This abandoned Milwaukee chair factory required five alarms to bring it under control in 1969. Milwaukee Journal *from James Haight.*

Hotel Holocausts

In the nineteenth century Milwaukee was known as the Cream City. The name was derived from the cream-colored bricks that paved the streets and composed the major buildings of the area. Cream City boasted that it had the "largest machine works, breweries and tanneries with the lowest death rate, the least crime and most home owners per capita." Yet such thriving industry and prosperity knew its dramas and tragedies.

The city had fine hotels. One of the finest had been the United States Hotel, but it was completely destroyed by fire in 1854. Recognizing the commercial necessity of fine accommodations in the business district of the community, civic leader and capitalist Daniel Newhall built a grand hotel, the Newhall House, to replace the ruins of the United States Hotel. It was a six-story structure made of the famous cream-colored bricks in the popular, but retrospectively quaint, "gingerbread" style of architecture. Although proclaimed as Wisconsin's largest and finest hotel, the Newhall House, for all its predicted prosperity, did not do well. It passed through a series of owners, each determined that he could make the three hundred room structure a success, but no one could make it profitable.

The hotel had its share of small fires, and its structural shortcomings were made public knowledge by editorializing newspaper reporters. Ample fire escapes were noticeably absent from the design as were enclosed interior stairways. When elevators were added in 1874, the elevator shafts were unprotected.

On January 10, 1880, the guests staying at the Newhall House had been aroused at approximately 3 a.m. by cries of "Fire!" Flames were discovered in the northern section of the hotel roof. It was believed that the fire started in the kitchen and burned through a partition into a ventilating shaft that pulled the fire up to the roof. Fortunately, some of the men were able to extinguish the fire with hotel apparatus. Two fire engines dispensed to the scene aided the cause from the street. Water damage was the primary factor in the $6,400 insurance settlement. Changing any part of the structure for safety's sake, however, was not suggested by anyone.

Local insurance agents later refused to insure the building because they considered it to be a firetrap. The inside woodwork was very dry and the partitions were not filled in with brick. And the large building provided only two fire escapes. Furthermore, because guests at the hotel had been considerably panicked in the 1880 blaze, the hotel staff was instructed not to arouse and alarm guests in "minor" fires.

On the night of January 9, 1883, exactly three years after the 1880 fire, the hotel was occupied by forty maids, laundresses and kitchen help and about three hundred guests, including the midget stage attraction General Tom Thumb, his wife and entourage. The night staff at the hotel consisted of a watchman, night clerk and elevator man.

At approximately 4:10 a.m. on the morning of January 10, the Newhall House was again on fire. The cause was never discovered but the folklore of the day determined it was arson. The fire appeared to have started near the elevator on the ground floor, and the unprotected elevator shaft served as a chimney to stoke the flames and proved to be an excellent means for transporting the flames upward to engulf all six floors. Alarm box 15, located near the scene, was pulled, and the crowds and the equipment began to arrive.

The fire was well beyond control as attempts were made to rescue the screaming, terrified inhabitants. The upper floors had such an intricate pattern of hallways that it was nearly impossible to escape from the maze. A description of the scene emphasized the hopeless panic and confusion: "The unfortunate

On January 10, 1883, Milwaukee's Newhall House burned in a fire of presumably incendiary orgin. The mazelike pattern of hallways on the upper floors made escape difficult and finally, in desperation, guests began to jump. The telegraph wires along the street entangled many in their leaps and hindered fire-fighting operations. State Historical Society of Wisconsin.

Firemen Herman Stauss and George Wells performed heroic rescues during the Newhall House fire of 1883. James Haight.

inmates were in many cases only aroused from their slumbers by the noise of flames and found their escape already cut off. Men, women and children rushed up and down the halls in the dense suffocating smoke, missing in their frantic efforts the stairways and windows leading to the fire-escapes." As the spectators gathered in the streets, the patrons gathered in the windows. A popular rhyme expressed it thus:

Milwaukee was excited as it never was before,
On learning that the fire bells all around
Were ringing to eternity a hundred souls or more,
And the Newhall House was burning to the ground.

With the fire at their backs, the guests began to jump. Telegraph wires surrounded the building and many became cut and entangled in them before meeting their deaths on the pavement. The wires also seriously detained rescue and fire-fighting efforts. Some people tried to jump into an outstretched canvas held by citizens, only to receive fatal injuries.

Hook and Ladder Company No. 1 was the first to arrive on the scene and promptly began fighting the blaze with the temperature four degrees below zero. Herman Stauss was a member of that company and recalled later that the hotel inhabitants were "shrieking and calling for help." Most of the servant girls on an upper floor did not even have a chance of escape and were killed. The extension ladder failed, but Herman Stauss and another fireman, George Wells, tried to rescue as many as possible by making a ladder bridge from an adjoining building to the burning hotel. The ladder spanned twenty feet above an alley. Stauss made several trips across the perilous bridge, saving sixteen girls. He became a hero and received tokens of appreciation from across the country.

Another fire fighter heroically led many persons from the hotel's fourth floor through the dense smoke, encouraging them and ultimately leading them to a fire escape. Later he lost his own life when a wall fell and brought down the wires with it. He was entangled in the wires, attempting to free himself when he was hit by a falling telegraph pole and buried in the ruins.

General Tom Thumb and his wife escaped injury but a member of their troupe was killed. A leading Milwaukee businessman and his wife jumped from the third floor and were fatally injured. In all, seventy-one people lost their lives at the Newhall House fire, the most deaths in a hotel fire until 1946.

The year 1946 was a disastrous year for hotel fires. On June 5, sixty-one people lost their lives at the LaSalle Hotel fire in Chicago. On June 9, nineteen people died at the Canfield Hotel in Dubuque, Iowa. On June 21, ten people lost their lives at the Baker Hotel in Dallas. But the worst hotel fire in the history of the country occurred at Atlanta, Georgia, on December 7, 1946.

The Winecoff Hotel was located in the heart of Atlanta's business district on the corner of Peachtree and Ellis streets. It contained 194 rooms and was well known in the Atlanta area. W. F. Winecoff had built the proud hotel in 1913. He had since retired but resided at the hotel. The building was of supposedly fireproof construction and hence did not have any fire escapes. "Fireproof construction" is a misnomer. The National Board of Fire Underwriters has since discontinued use of the term, supplanting it with the term "fire resistive." Fireproof merely meant that the framework of a building will remain sound after a fire. It said nothing of the contents and, unfortunately, people are not fireproof.

The building was fifteen stories tall with the floors numbered consecutively except for the number thirteen which was eliminated from the numbering system. The structure was protected by a shielded

Firemen in New York City organized their own regiment, The First Fire Zouaves, at the outbreak of the Civil War. On May 9, 1861, they were bivouacked in the Capitol in Washington, D. C., when the famous Willard Hotel caught fire. Without being asked, they rushed to attend successfully to the blaze with some spectacular feats of fire fighting.

The Adams Hotel (above), built in 1896, was the pride of Phoenix, Arizona, in 1910. On May 17 of that year a spectacular fire broke out. Fortunately all inside escaped, so firemen concentrated on containing the blaze. Men several hundred feet away protected themselves from the immense heat with blankets (opposite). Both Arizona Photographic Associates.

steel frame, and the roof and floors were of concrete on tile filler. The exterior was composed of twelve-inch-thick brick panel and inside partitions were constructed of tile plastered on both sides, insuring structural stability. On the other hand, the walls and corridors were covered with painted burlap from the wood baseboard to the rail, above which they were papered. Corridor floors had wall-to-wall carpet on felt padding. Doors to rooms were of light panel wood, with wood frames and transoms. The rooms were wallpapered, some with as many as five thicknesses of wallpaper, and ceilings were painted. While a few of the guest room windows had the luxury of wooden venetian blinds, most of the windows were covered with ordinary cloth draperies. While the building itself was indestructible, apparently little thought had been given to the flammability of the contents, which were not. A kitchen stove, for example, while it contains flame for controlled use

and function, still can burn the flesh and in all probability would not be a good place to sleep.

The building design also included many openings, mostly vertical, such as ventilating shafts, to aid in the building's function to serve its guests and make them comfortable. These openings also had a hidden use: In the event of fire, they would serve as chimneys and fans to draw the oxygen-seeking flames onto all fifteen floors. The Winecoff Hotel was also equipped with open transoms above guest-room doors, and an open stairway, the single means of escape in the event of an emergency.

The two elevator shafts were centrally located while the single stairway was also in the center of the floor plan. The stairs began on each floor as a single staircase then branched off into opposite directions halfway up, each stairway leading to the two long corridors that ran parallel to each other. Since the elevator shafts were enclosed with fire resistive

materials, a fire, should it occur, would probably proceed up the staircase feeding on the burlap, wallpaper and combustible woodwork.

On the morning of December 7, 1946, the Winecoff Hotel was filled nearly to capacity with about three hundred guests on the hotel register. At the corner of Peachtree Street and Carnegie Way, a short block from the hotel, "Gone With the Wind" was playing at a theater. Handbills in the hotel lobby advertised "Three Little Girls in Blue" and the Walt Disney fantasy, "Song of the South," playing at nearby movie theaters. Around 3 a.m., when most of the guests were soundly sleeping, the fireproof hotel caught fire. The fire went down into the official records as being of "undetermined origin." It started on the third floor and experts later speculated that it began in a carelessly stored mattress in the corridor.

The night crew at the hotel consisted of a night desk clerk, a bellman, a night building engineer, an elevator operator, a night maid and a cleaning woman. Shortly after three o'clock the bellman was called to Room 510 to deliver some ice. The night building engineer, whose habit it was to tour the building at intervals to certify that everything was in order, accompanied the bellman to the fifth floor.

Both were invited into the room by the hotel guest and they accepted the invitation. At about the same time the night elevator operator took some guests to an upper floor, probably the tenth. On her

Most victims in the Winecoff Hotel fire were on upper floors (above) as tongues of flame shot from windows. At 9 a.m. firemen began their search for bodies (left). Both Wide World.

Chicago's LaSalle Hotel burned June 5, 1946, killing sixty-one people. The fire began behind a false wall in the cocktail lounge and roared through the lobby, consuming the combustible paneling. Afterwards, only charred ruins remained of the lobby and mezzanine. Wide World.

way down, she smelled smoke somewhere between the third and sixth floors and reported it to the night clerk when she reached the lobby. The night clerk then instructed her to go to the fifth floor to locate the engineer and bellman. As she ascended, the night clerk ran up the stairway to the mezzanine where he saw the reflection of flames from the third floor and ran back to the lobby to telephone the fire department. He then began to phone individual rooms to warn the guests. The elevator operator was undoubtedly shaken by the events and was unable to locate the bellman and engineer. She returned the elevator car to the basement and ran up the stairs to the lobby. The elevator girl later reported that she saw flames as she left the fifth floor. If this is so, the fire was making swift progress up the staircase.

The fire department responded immediately. The night clerk turned in the alarm by telephone at 3:42, and a fire company only a block away arrived within thirty seconds. Though the fire department arrived with haste, when they arrived the third to fifth floors were engulfed in flames, and guests were already jumping out the windows. Stairways and hallways above the third floor were filled with smoke and choking gases. The fire spread so rapidly due to the natural chimney provided by the open staircase that escape from the upper floors was impossible.

Modern building laws require two safe, readily accessible exits from every corridor, public space or service area. The Winecoff Hotel had only one principle means of escape — the open stairway. This was 1946, a year that had already seen its share of tragic hotel fires, and just eight years previous Atlanta had experienced its own tragedy. On May 16, 1938, the Terminal Hotel in Atlanta had burned, killing thirty-eight people. Still the Winecoff Hotel was allowed to stand, without a single fire escape. After the fire, city authorities expressed amazement that the hotel provided no emergency means of exit. It was reported that the hotel had recently been inspected by the fire marshal and had measured up to safety standards. Governor Ellis Arnall wanted a full investigation of the disaster. He said, "This is a great tragedy. The public is being defrauded when a hotel is advertised as 'fireproof' but really isn't. Responsible agencies should prohibit the use of the word 'fireproof' when a hotel is not really fireproof as the Winecoff obviously was not."

Walls crumbled to the street as fire raced through the Normandie Hotel in Philadelphia on January 8, 1968. The hotel was primarily a residence for elderly people. Though about 325 were evacuated, no one was hurt. Wide World.

As people implored rescuers for help from their windows the spectators gathered in the street to watch. People began to lower themselves from one floor to another with tied bed clothing. Many fell to their deaths when the knotted sheets broke or were burned, and some died when they could hold on no longer. Fire fighters placed ladders onto the building, but many could not wait to be rescued and jumped onto the street. Some were saved by the life nets held by firemen, and many were able to climb down the ladders provided. Rescue by ladder was limited to the length of the ladder, but there were ladders that reached the sixth, eighth and tenth floors. Rescue was attempted from window to window and people attempted to lower themselves to the ladder. One man tried to reach a ladder and as he swung down and was suspended, two other people fell onto him from the floors above and all three plunged to their deaths. A fireman was taking a woman down a ladder when another woman fell onto them. The three toppled to the hotel marquee; the two women were killed, the fireman seriously injured.

After the fire, it was noted that more than half of the rooms had the transoms open above the door. The open windows and open transoms served to draft and fan the fire. The fire companies on the scene turned in three subsequent alarms calling for the entire department; it took three hours to bring the roaring inferno under control and an additional three hours to put it out. Every available piece of equipment was there. Help arrived from the suburbs: some companies to help fight the fire, some to maintain fire protection for the rest of the city.

Many were burned to death; many jumped; many tried to escape into halls and were overcome by smoke and gas and asphyxiated. One woman was said to have hurled two small children from a window and then leaped to her own death. A man died when he missed the life net by only inches.

In this fire, too, just as in the Newhall House blaze, many were saved by ladder bridges across an alley between the burning hotel and the neighboring Mortgage Guarantee Building. Major General P. W. Baade of Washington, D.C., and his wife were rescued in this manner. The major had commanded the 35th Division in World War II. He said that facing bullets on the battlefield could not compare to the hopelessness and fear of being trapped by the tongues of a hot fire.

The bellman and engineer were rescued by ladder. The bellman's testimony indicated that they had not been in Room 510 for more than two minutes. When they turned to enter the hallway, they were trapped by flames immediately.

One couple crawled along a fourteenth-floor ledge to another couple's room. When they reached the other couple's room they stacked mattresses against the door and kept them wet. They survived.

The story from a room on the eleventh floor did not end as well. A mother gathered her three children in her arms and held them close — all died.

At 9 a.m. on December 7 the room-to-room search for bodies began. On the outside the Winecoff Hotel was draped with bedsheets and blankets hastily tied and now hanging limp and lifeless from the windows. The walls in many rooms were burned to the tile base. Most doors and window frames were burned. Mirrors were shattered and windows were smashed. There was much evidence of heat in excess of 1,500 degrees Fahrenheit. Light bulbs were fused, the heavy metal elevator doors were twisted, telephones melted. In some rooms only the bedsprings remained; the rest of the furnishings were wholly consumed. Yet the fireproof hotel was still structurally sound.

It was the worst hotel fire in American history. At its conclusion 119 people died and an additional 91 were injured. Very few people escaped unharmed. Among the dead was W. F. Winecoff, seventy, builder of Atlanta's famous fireproof hotel.

School Bells and Fire Bells

The Lake View Elementary School in Collinwood, Ohio, ten miles east of downtown Cleveland, and today a section of that city, was built in 1902, five years after John Dewey, the famous American educational philosopher, wrote *My Pedagogic Creed* and three years after his *School and Society* was published. One of his principles was: "The school is not a preparation for life; it is life." On March 4, 1908, Lake View School — "life" for over three hundred young students — caught fire. Dewey's lofty principles were lost forever to the 175 who died within its walls; the school, while maintaining conditions of life, was not prepared for death.

The schoolhouse was an imposing brick structure consisting of two stories and an attic. Overcrowding at the school led to the use of the attic for classes, and it was called the third floor. The school population consisted of students from six to fourteen years of age, with the six- to eight-year-old children occupying the attic. Preparing students for their roles in a society based on democratic participation — another Dewey philosophy — took place in nine classrooms, each with its own teacher.

The building itself was later described as being of inferior construction. It was made mostly of wood with brick walls. Little regard was paid to construction details that would save lives should there ever be a fire in the school. It was hoped that frequent fire drills would compensate for the building's shortcomings. There was only one fire escape, and its use was not taught. Wing partitions narrowed corridors by three feet, making traffic flow difficult, especially in an emergency situation. In addition, according to a newspaper description after the disaster, "the halls and stairways were enclosed between interior brick walls, forming a huge flue, through which the flames shot up with great rapidity."

Drills were regarded as the primary form of fire precaution. In many areas of the country, in fact, fire drills were mandatory. A believer in fire drills as the primary life-saving means in schools asserted that "where they are carried out frequently and at unexpected hours the children become so well-disciplined that they act in a real fire precisely as in a practice drill and often get out of a building without even knowing that there is a fire." The school in Collinwood held frequent fire drills; however, when there was an actual fire, there was panic. All discipline was lost in the struggle to flee the burning building. A *New York Tribune* account said, "Penned in narrow hallways, jammed up against doors that only opened inward, between 160 and 170 children in the suburb of North Collinwood were killed today by fire, smoke and beneath the heels of their panic-stricken playmates." The children, who had been "well-disciplined" in the fire drills, could not be controlled in the actual fire.

Fred Herter was the janitor at the school. It was a warm, springlike day and the fire in the furnace was lower than usual. Herter, while sweeping in the basement and trying to decide whether the fire was too low and whether to open the vents, said that he "saw a wisp of smoke curling out from beneath the front stairway. I ran to the fire alarm," he continued, "and pulled the gong that sounded throughout the building. Then I ran first to the front and then to the rear doors. I can't remember what happened next, except that I saw the flames shooting all around and the little children running down through them, screaming."

The building contained two stairways: One led to the front, the other led to the rear. Flames first appeared at a closet that contained lime and sawdust near the front exit. It was later believed that a malfunction in the furnace had started the fire, since the furnace was located in the basement directly beneath the front exit. The corridor was almost instantly filled with smoke and flame. As soon as the teachers were made aware of the fire by the sounding of the gong, students were lined up according to fire drill procedure. However, fire drills had usually led to the front door, and as the students approached this door, flames were already lashing out from the area, blocking the exit. The students, untrained to leave by

Ominous smoke curled behind windows in the Lake View School at Collinwood, Ohio, on March 8, 1908. Volunteer firemen focused their weak stream through a window, but the damage had already been done: Though the school held frequent fire drills, 175 people, mostly students, died. Cleveland Plain Dealer.

any other exit, became frightened and stumbled over each other to reach the front door before the flames made it impassable.

Following the fire, parents made angry accusations that the front door was locked. This was unimportant, however, since the congestion of children in panic — fighting for their lives, stumbling, being trampled and falling over their classmates — "effectually barred the way" out of the front door, as one description of the scene was reported.

Children from the second floor raced wildly to the foot of the stairs with the flames closing in. They instinctively regarded the door as the only chance to save their lives; firemen later found "a tightly packed mass of children . . . piled against it." After that first rush to the front door and the wall of bodies that was suddenly mortared there, no one using the first flight of stairs could escape. Those descending in a some-

what orderly fashion from the second floor saw the destruction, turned to go back to the second floor and were trapped by the third floor occupants now crowding the stairway to seek the only emergency exit they knew. More panic resulted; two hundred students fought for their lives. The majority of dead were found in this area, and all who were caught at the foot of the first stairway died; none were able to give the details of what happened there. But the panic was the result of the neglect that caused the deaths in the first place. As one dispatch said: "After the flames had died away . . . a huge heap of little bodies, burned by the fire, and trampled into things of horror, told the tale as well as anybody needs to know it."

The school contained 310 to 325 pupils, but only eighty were able to leave the building unhurt. The school janitor lost three children of his own in the

Students moved orderly toward the front door (above) but flames lunged at them and passage through the front door was impossible. Children panicked and were found packed against the door. Cleveland Press. Makeshift ambulances (opposite) wait to claim the victims. The fire escape in the back saved nearly everyone from the third floor. Cleveland Plain Dealer.

blaze. He was watching the hundreds attempting to escape: "Some fell at the rear entrance and others stumbled over them. I saw my little Helen among them. I tried to pull her out, but the flames drove me back. I had to leave my little child to die." Herter himself was badly burned around the head. Despite his great personal loss, Herter was blamed by grieving parents for the fire. After all, the fire began in the basement and that was the janitor's territory. One irate father even tried to kill Herter and the school janitor had to be placed under police guard.

Railroad workers at the nearby Lake Shore railway shop rushed to the school upon hearing the alarm. Their thoughts were to rescue as many as they could and their first action was to force open the front door. When the door was finally opened the workers saw the wall of flames that had formed across it. But they were too late to save any of the students — most of them were dead by the time the doors opened.

Subsequent investigation definitely showed that the rear door was locked. Miss Catharine Weiler, a teacher on the second floor, lost her life at that door after she attempted to save the lives of her students. When the alarm was sounded she routinely lined up her pupils for what she thought to be another fire drill. The fire had been allowed to progress rapidly and it was soon apparent to the students that the school was burning fiercely and that there was a good chance of being trapped and burned to death. At that point, Miss Weiler lost all disciplinary control over her students. "Children in their frenzy lunged into the struggling mass ahead of them" as Miss Weiler tried to reassure them. Her body was found buried beneath the bodies of those she taught.

Mrs. Clark Sprung lived in one of the few residences near the school. Her son, Alvon, age seven, was a second-grade student at Lake View School. She arrived at the building when the first floor was engulfed in flames. Seeing her son at a first floor window, begging for help, Mrs. Sprung quickly dashed across the street and procured a ladder from a neighbor. Upon climbing the ladder she was able to catch hold of her son by his hair but, a press dispatch sadly noted, "it burned off in her hands and the lad fell back into the flames."

Walter C. Kelley, the sports editor for the *Cleveland Leader,* lost two children in the blaze and was the first to angrily report that the doors to the building were locked. His wife ran from their nearby home to the burning structure. The front of the building was a mass of flames but she desperately hoped that she could save her children at the rear door. Her attempt to save lives was driven by "the screams of the fighting and dying children which reached her from the deathtrap at the foot of the first flight of stairs and behind the door." She ran to the back door where she was joined by a man, probably a railroad worker. They "tugged and pulled frantically" at the door, but it would not move and they did not

burned, but he could not be persuaded to leave — until he saw that his little girl was dead.

Marie Witman heroically ran through the hallways and grasped the hand of her little brother and then helped him through a window. Both nearly suffocated. Miss Ethel Rose, a first-floor teacher, managed to lead all but three of her charges to safety, carrying two small children in her arms. Miss Laura Bodey, the only teacher on the third floor, methodically lined her students up and with the normal regimentation, marched them the customary way to the second floor. She saw the flames rushing up the stairway and turned her students back, still maintaining discipline. After they reached the third floor, she broke a window with a chair and climbed onto the fire escape. She helped her students down, one by one. Five students had broken from the line in the confusion, and were not seen alive again.

Another teacher was carried by the panic toward the rear exit. She fell at the bottom of the staircase. Panic-driven pupils fell on top of her and the weight of the bodies made it impossible to get up. However, she was rescued in time to save her life. The principal and two other teachers abandoned their students and escaped through a rear window. This action was defended in contemporary accounts of the tragedy since it was contended that "they remained with the panic-stricken children until they could do no more for them and then sought their own safety."

It was estimated that it took less than five minutes for available means of escape to be cut off. After the fire was allowed to progress for half an hour, blackened walls and charred bodies were all that remained. The fire ate through the wooden cross supports on the first floor and licked its way upward until all three floors crashed into the basement.

Fire-fighting efforts were such that the fire was forced to burn itself out. Collinwood's eight thousand people were protected by a volunteer fire department whose equipment consisted of two inefficient fire engines. The department did not even have a ladder that could reach the third floor of Lake View School.

have the equipment to break it down. They desperately worked to save some children, but soon the pleading voices within were silent. Before all were completely overtaken by the flames, Mrs. Kelley and the man smashed a few windows and saved some children. Kelley angrily charged that they could have saved many more if the door had not been locked.

Crowds gathered on the school lawn. Parents spurred only by the thought of saving their children battled with police to get into the flame-swept structure. Wallace Upton reached the building shortly after the front door fell in. The crowds were then able to see the devastation that had taken place inside. The fire now had a fresh supply of oxygen through the opened front door and gained new momentum. Upton saw his ten-year-old daughter before him — crushed, burned, trampled, but alive. The flames were quickly approaching, and Upton pulled at her with all his strength. His clothing was burned from him, his face and hands were badly

The crew was inexperienced in fighting such a blaze and one report accused them of arriving at the scene late. It was soon discovered that the two engines the department did have were ineffective since the water pressure was too low to get a good stream on the fire.

Firemen attempted rescue operations despite their limitations. When the locked rear door was finally broken down, the anxious women outside saw before them what was described as a "mass of white faces and struggling bodies." Flames overtook the children while the crowd helplessly watched, causing many of the women spectators to faint. As soon as they were able to enter, the firemen searched for life. No one was taken alive after the floors collapsed. In addition to those burned or asphyxiated, little Mary Ridgeway, Anna Roth and Gertrude Davis were instantly killed in their leap from the third floor to the ground. The fire was finally completely out in three hours.

Others helped the firemen in the rescue operation. The railroad company donated their nearby building as a temporary morgue and a line of rescuers formed at the schoolhouse, backed by six ambulances. A large crowd, nearly the whole town, formed around the burned building during the rescue operations. Parents waited outside the school all night for some word of their children's fate. One father and mother stumbled around the ruins for hours looking for their daughter and were still on the scene the next morning waiting for word when Helen Marks threw her arms around her father's neck as he was prodding about the burnt remains. She had been sent to school as usual that Wednesday but decided that such a beautiful day was made for better things than school. She played hookey and was visiting an aunt in the country while many of her classmates were crying for help in a burning building.

Pledges and money poured into Collinwood after the tragedy. Fire drills were suddenly held all over the country, and school authorities suddenly found themselves on the defensive, expounding on the virtues of their various school districts and insisting that their schools were safe from fire, that "it could never happen here because we have fire drills." Meanwhile, bodies of children were being identified at the morgue by clothing and trinkets, and funerals were being held for students who lost their lives in a school that also held fire drills.

On December 1, 1958, in Chicago, between 1,200 and 1,300 students were settling down for the last hour of classes at Our Lady of the Angels parochial school. A little more than thirty-five minutes before the 3:00 p.m. dismissal, a fire started at the bottom of a stairway in the north wing of the school. While the exact cause was never determined, it was guessed that a student sneaking a cigarette may have caused the fire that killed ninety pupils and three nuns.

Two students on an errand returned to their classroom saying that they smelled smoke. The teacher took them seriously and hurriedly consulted with a neighboring teacher — both decided to evacuate their students. The rest of the school was not, at this time, alerted to the fire. The two classes left the building — one class using the fire escape, the other an inside staircase — and reported to the church on the same grounds. The janitor entered the building and noticed that it was on fire; he told the parish housekeeper to call the fire department. It was suggested later that she may have delayed calling since an alarm was not received until 2:42.

The building fire alarm was also not sounded until 2:42. This alarm was manually operated and was not connected to the fire department. Yet it was the first indication to the occupants in most of the rooms that the building was on fire. Smoke was noted to be on the second floor, though the fire had probably not

Word of the Our Lady of the Angels fire spread quickly and huge crowds watched and waited as fire fighters continued their efforts (opposite). Chicago Daily News *photo by Mat Anderson. Only charred debris and a collapsed roof (below) remain of a schoolroom in which nine pupils and a teacher perished.* Wide World.

A roof collapses in a spectacular burst of fire in one of Houston's oldest elementary schools. At least ten firemen were injured in the recent four-alarm blaze. Wide World.

reached that level. But three minutes later children could be seen calling for help from second-story windows. Since it was the imploring, pleading children that most concerned the firemen upon their arrival, the fire eventually progressed to such a state that it burned off a large portion of the roof.

Occupants of the first floor safely evacuated in fire drill formation. The situation was more tense and difficult on the second floor which was now filled with thick clouds of smoke. There were many instances of heroism and tragedy on the second floor. In one classroom the children were gripped with fear and refused to leave. The teacher instructed them to crawl to the staircase and she pushed them down and out, saving them. In another classroom, students attempted to leave but were turned back by smoke. Some jumped and some managed to crawl down short ladders to safety.

Math was being taught in another room when the tragedy struck. The teacher here, in addition to being heroic, was quick-thinking. She ordered students to pile books around the doors where the smoke could seep in. She told them to put desks in front of the doors to keep out the smoke and the tremendous amount of heat that was beginning to come in. The alarm had not yet been sounded and she told the class to chant a message in unison that said the school was on fire. By keeping students constructively occupied and by quick actions that increased their confidence in her, all remained calm and awaited rescue. Some safely jumped to a stairway a few feet below, while other waited at windows for fire department ladders. When all her students had been rescued, the teacher descended on the ladder. Though this room was heavily damaged by smoke and fire, only one pupil died. The students in the classroom across the hall were not as fortunate — the teacher and twenty-nine students died; many were burned to death.

Before the fire department arrived, rescue efforts were attempted by school personnel and passersby. One person risked his life to check and make sure that everyone had been evacuated from the first floor; someone else was seen carrying four children; still another concerned citizen guided a group of children to an exit. It was estimated that nuns, lay teachers, priests, janitors and passersby saved one thousand of the children.

Upon its arrival, the fire department worked heroically to save an additional estimated 160 lives. It took five alarms to control the blaze and effectively perform rescue operations. Firemen groped their way through smoke and fire seeking children that might be trapped. They fought their way into a classroom only to find twenty-four students dead at their desks. Hallways and rooms were filled with smoke and gases and yet the firemen remained, looking into every room for signs of life. They found many groups of children still alive. One twelve-year-old girl said after the fire: "Some of the boys jumped out of the window. When we looked down we saw them lying still on the ground. We stayed and the firemen saved us."

The building had no automatic sprinkler system, was not of sound construction and the stairways were not enclosed, which served to draw the fire up and feed it. And Our Lady of the Angels, a place of prayer and learning, burned, taking ninety-three lives with it.

Milwaukee's old Eighteenth Street Elementary School burns brightly during a 1973 snow storm. The school had been permanently closed down only months earlier, thus preventing what might have been a tragic loss of life. Milwaukee Journal.

Theaters:

The Show Did Not Go On

Kate Claxton, a famous actress who was delighting the audiences with her performance as the heroine in *The Two Orphans*, was on the stage of the Brooklyn Theater in New York, which had been transformed into a boathouse for the last scene. It was eleven o'clock on the night of December 5, 1876, and soon the capacity audience would be discussing the performance as they slowly filed out. A canvas backdrop broke from its fastenings and dangerously hung over one of the border lights in the center of the stage. One of the stagehands quickly responded and hoisted the piece upward. However, he raised the canvas too quickly and the resulting rush of air caused the smoldering backdrop to suddenly burst into flame. A roof ventilator fanned the flames onto other backdrops. The stage manager ordered two stagehands to extinguish the fire, but they were unable to do so and soon the entire stage scenery was ablaze.

Kate Claxton described the scene:

> At the beginning of the last act just as the curtains went up, I heard a rumbling noise on the stage, and two minutes after I saw flames. The fire seemed to be all on the stage. Mrs. Farren, myself, Mr. Studley and Mr. Murdock were on the stage at this time. We four remained there and endeavored as best we could to quiet the audience and prevent a panic. I said to the people "Be quiet, we are between you and the fire. The front door is open and the passages are clear." Not one of the audience jumped on the stage. The flames were then coming down on us. I ran out and jumped over several people. Mr. H. S. Murdock, after endeavoring to calm the fears of a panic-stricken people, went to his dressing room to get his clothing, and must have been suffocated.

The fire marshal conducted an investigation after the tragedy and a witness reported to him that ". . . fire commenced to fall on the stage. When I saw it at first, the pieces were dropping on the stage. As soon as the cry of fire was raised, people rushed to the door, but when told by the actors to be quiet, some sat down again a few moments. Then the fire

became visible and was followed by a dreadful panic." Many of the actors and actresses were able to escape by backstage routes to Johnson Street. The main exit for the audience was on Washington Street, and this immediately was choked with terrified people. One witness said that "the people were panic-stricken and were falling on each other."

In the surge from the upper galleries to the street a woman caught her leg in a railing on a landing. The crowd tripped and fell over her, and soon there was a large pile of struggling people. Police immediately arrived from the station house next door but found the audience packed so tightly that few could be removed. Most of the dead were burned or trampled beyond recognition. Firemen began to remove bodies on the morning of December 6 and found them packed in tight rows beneath the rubble, indicating the great pressure they had been subject to. Many victims were found in the basement, since all the floors had collapsed and crashed downward. Washington Street was lined with horses and wagons ready to remove bodies to the morgue.

The ruins still smoldered in the late afternoon of December 6; flags flew at half-mast and other Brooklyn theaters were draped in mourning. Actors Claude Burroughs and Harry S. Murdock had lost their lives. The fire had vented its fury on the theater patrons and 295 people were either burned, trampled or suffocated.

About twenty-five years after the Brooklyn fire, Chicago built a large, modern, "fireproof" theater patterned after the *Opera Comique* in Paris. The Iroquois Theater opened its doors in November of 1903. The following month, during Christmas week, many families attended the special afternoon performance of the light opera, "Mr. Bluebeard," which featured the theater company's popular comedian, Eddie Foy. Children were on Christmas vacations from school and parents deemed it an appropriate time to take the children on an outing to the theater. The theater was advertised as the most beautiful in America and was embellished with luxurious details. The seats in the "fireproof" theater, however, were

Harry S. Murdock, Kate Claxton and two others were on stage during the Brooklyn Theatre fire of 1876. Kate Claxton pleaded, "Be quiet, we are between you and the fire," in a futile attempt to prevent panic.

wooden and stuffed with hemp. The building was only five weeks old, and the precautionary fire equipment, "just in case" a fire should ever break out, had not yet been installed.

The theater was tightly packed with men, women and children — many entire families — the afternoon of December 30. The "Standing Room Only" sign was up outside, but many people did not mind standing for a good performance. The stage was partially darkened for a moonlight scene, and a powerful spotlight followed the actors. Part of the stage drapery came into close contact with the light and caught fire. Some stagehands saw the fire and tried to smother it with their hands and other primitive means; however, the fire was more than they could handle. The audience was made aware of the fire as a flaming piece of drapery swung across the

stage. Performers, aware instinctively of the consequences, played on in show business tradition.

A fire alarm was turned in. Eddie Foy, waiting for his cue backstage, heard the commotion and rushed onto the stage. He ordered the orchestra to keep playing — "Play, play and keep playing!" — while, amidst great applause and appreciation from the admiring audience, he attempted to reason with the people. He clowned a bit to relax them and then spoke seriously, telling them not to stampede. The comedian ordered that the asbestos curtain be dropped to prevent the flames from going beyond the stage. The curtain was dropped, but not all the way. The audience began to file out as the band still gaily played. Eddie Foy, determined to prevent a panic, danced and clowned as the audience was leaving. The fire, however, was still gaining momentum behind the curtain, out of sight of the audience. Suddenly the scenery crumbled in flames, and the fire lashed out from beneath the asbestos curtain. It was at this time that some reports of the tragedy indicate that the theater was plunged into darkness.

Flames lunged into the balconies and galleries, and people screamed and panicked. Their only motive was self-preservation — some jumped from balconies into the stampeding crowd below, while others were crushed in the struggle. The fire department arrived but the firemen were unable to get into some of the upper galleries to save any lives since the entrances were blocked tight with bodies. "Flames," according to a contemporary report, "caught the people before they realized the full extent of the danger." Many of the patrons seated in some of the balconies were so quickly overcome that they died without ever leaving their seats. People crowded onto the fire escape from the doomed balconies but the fire was shooting out from windows below them and that route was useless. All were out of range of aerial ladders and still more people crowded onto the platform. Painters in the next building provided ladder-bridges across the alley for the trapped people. "Ladders, planks, ropes, poles, everything that could possibly be secured to assist these poor creatures in their battle for life was rigged and turned into bridges, but few got across alive," a newspaper reported.

Using an adjacent alley, firemen work to put out the Iroquois blaze (above). The words, "Absolutely fireproof," ironically stand out on a charred theater program (right). Both Chicago Historical Society.

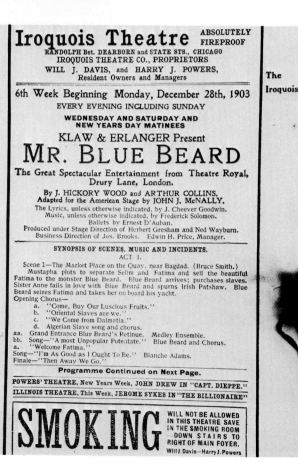

Firemen were on the scene fifteen minutes after the alarm was turned in, attempting to extinguish the blaze, to rescue people, and struggling "to do what they could." After a fireman discovered that the bodies in some of the balconies were piled high and jammed, he quickly reported to Fire Chief William Musham that if any were to be rescued, it would have to be done promptly. The chief then ordered the fire fighters to abandon efforts to extinguish the fire and concentrate on rescue operations.

Hospitals sent doctors and nurses to the scene, and news of the holocaust spread rapidly. People came from every section of the city to seek their loved ones

*On December 27, 1886, the Temple Theatre in Philadelphia burned despite daring
and determined fire-fighting efforts. Firemen quickly trained several strong streams on the
blaze and attempted to gain an advantage of height by scaling the many ladders set up.*

Fire consumed the Nautilus movie theater (above) in Long Beach, New York, in 1972 despite firemen's efforts to save it. Moments later, the marquee crashed in a shower of sparks (opposite). Both Robert Preudhomme.

who were known to be at the theater for the matinee. Within ten minutes after firemen began to remove bodies, a dozen stores in the area had been converted into temporary morgues. The owner of the theater blamed the loss of life on the confusion and panic. He claimed that if everyone had remained seated at the first cry of "Fire!" no lives would have been lost. Firemen, however, found many patrons who remained seated — and died. Of an over-capacity crowd

of nearly two thousand, 602 lost their lives. The stage and parts of the audience area were destroyed within forty short minutes.

In the investigation after the fire, it was discovered that many flammable materials had been included in a theater advertised as "absolutely fireproof." There were no exit signs and many people had followed paths that they thought led to freedom only to find they led to dead ends. Regulations were passed after this fire to provide for lit exit signs in all theaters, and for such other fire precautions as sprinkler systems, extinguishers and fire-resistive scenery.

The Iroquois fire was the worst American theater disaster in history. A report said, "The classic outlines of the theater, the beautiful plush hangings, the arch windows with their stained glass, the stately pillars, became a morgue five minutes after the first little ribbon of flame made its way along the stage."

Government-Sponsored Firetraps

O. Henry, the American author, needed money. His wife had tuberculosis and required extensive — and expensive — treatment, and he wanted to start a literary magazine. He was conveniently employed by a bank and managed to "borrow" a little money now and then. When discovered, he ran away but returned in 1897 to be at his wife's deathbed. Convicted of embezzlement and sentenced to five years in the Ohio State Penitentiary, he served three of the five years from 1898 to 1901. It was here that he gathered much of the material for his brilliant short stories. Other lives of inmates in the history of the penitentiary did not end as happily.

A visitor strolling through the prison yard at the Ohio State Penitentiary at Columbus in 1930 could not help but catch a chill from the atmosphere of dejection there. A penal reform organization had proclaimed that the prison was one of the worst in the country and remained untouched by the spirit of social reform that had motivated changes in prisons in other states. The fortress was located in the middle of a large city, whose everyday sounds of living and freedom filtered in through the ancient rock walls. Life on the inside of the walls consisted of weary routine — menial and meaningless labor for the incarcerated at the rate of five cents an hour. The prisoners were known to contribute to progress, however. After Ohio abandoned death by hanging, a prisoner, Charles Justice, helped significantly in the building of the electric chair. Ironically, Justice was the first to occupy the chair when the current was finally turned on.

But the worst part of the deplorable conditions at the prison was the overcrowding. Often, those waiting on death row were crowded into cells which contained the most objectionable conditions. Much of the prison's foundation dated to the early nineteenth century, and the bleak, twenty-three-acre site had been designed to hold a maximum of 1,500 prisoners. In April of 1930 it held nearly 4,300.

Discontent became part of the air that the prisoners breathed. Several times, escape attempts were made. Prisoners, whether detained by the state for murder or car theft, wanted only to be let out of the "cage," as a prisoner called it. Once, as the huge gates opened to let visitors in, prisoners managed to storm the entrance and thirteen set themselves free. Warden Preston Thomas, a capable but "hard-grained" administrator, promptly felled four of the escapees with buckshot from his own gun. Another was apprehended on a side street while the others were returned to the penitentiary by a plainclothesman.

Life in the prison went on as bitter resentment on both sides grew. Informers to the warden told of elaborate plots to escape by either digging beneath the walls or by setting fire to the prison. The warden found that not all the threats were fabrications. As time went on, the reports of escape plans became more convincing. One prisoner allegedly confessed to the warden that he had been stealing candles from the soap shop and handing them over to prisoners who also had oil and gasoline in their possession.

Many prisoners worked on cell block construction, which was at last being provided to ease the overcrowding. Scaffolding had been erected next to the six-story tiers of cells. On April 21, 1930, between 5:20 and 5:50 p.m., a fire began in the scaffolding. Smoke ominously curled from the construction area and a fire alarm was turned in at 6:00. The men had just been returned to their cells from the evening meal and locked in for the night. Some were engaged in games of checkers, others munched after-dinner candy bars.

When the first sinister puff of smoke curled into the cells, the prisoners were relaxed and unprepared. But a strong wind had driven the flames to the tiers of occupied cells. The flames rushed in and soon the men were screaming and pounding on their cells, calling to the guards to unlock the heavy cell doors. Their agonizing cries echoed throughout the prison — and throughout the Columbus neighborhood that surrounded it.

There was confusion among the guards as to what to do. Should they free the prisoners, or was the fire

Firemen tackled the 1930 Ohio State Penitentiary fire from the prison yard before moving into the cell blocks. Over four thousand convicts occupied the prison at the time. It was suspected that inmates set the fire in an escape attempt, but the actual cause was never determined. Wide World.

serious enough to risk the possible escape of the inmates? At the beginning of the fire, no one was yet alarmed since it was generally believed that the prison was fireproof. And, too, there was the matter of locating the keys: One guard possessed the magic keys that could have saved the prisoners' lives in the fire area, but he could not be immediately located. The roof of the prison was extremely flammable and was seriously threatening the men in the upper tiers. When he finally arrived at the scene, the guard with the keys claimed that his superior ordered him to keep the cells locked. After the fire, his superior denied giving such orders.

Orders or not, other guards wrested the keys away from the adamant guard and proceeded to unlock the cells in the lower tiers. One guard was reported to have hastily run down the corridor unlocking cells until he was overcome by smoke. He then relayed the keys to a prisoner who just as heroically freed the men. Prisoners were corralled into the prison yard in the confusion and placed under heavy guard so that they could not escape. The upper tiers were now a mass of flames but many men were set loose from the lower tiers. In four or five minutes the screaming in the upper tiers no longer existed — the fire had already claimed its victims. Due to the lack of efficiency in locating the keys and unlocking the cells, the men died after flames reached the roof. By the time the cells were ordered unlocked, the upper

two tiers were engulfed in flames. The fire department arrived within two minutes after the alarm was turned in but could do nothing in the confusion. Near-riot conditions existed as guards and, later, militia were overcautious in preventing escape and the prisoners were belligerent and vociferous at being allowed to barely escape in time. The National Guard cordoned off the prison.

All the guards survived, but 320 prisoners died, many of them within days of completing long sentences. It was suggested that all could have been saved if the guards had unlocked the cells at the first sign of fire. One of the guards said: "I saw faces at the windows wreathed in smoke that poured through the broken glass. With others I tried to get to them but we could not move the bars. Soon flames broke into the cell-room and the convicts dropped to the floor. They were literally burned alive before our eyes." The mad confusion that reigned in the prison after the fire began was largely due to the fact that no training, instructions or fire drills had been given as fire precautions.

The prisoners themselves risked their lives. A convicted gunman carried twelve men to safety and later collapsed and died; a notorious Cleveland bank robber performed rescue operations until he, too, was overcome by smoke. The two fire department units first on the scene claimed that attempts were made by several prisoners to cut their hose. Streams of water from the prison yard were of no use. Firemen were able to contain the fire briefly at one point but the strong wind hampered their efforts. Convicts, prison guards and firemen worked side by side to get into the doomed cell blocks. Firemen used acetylene torches to drill the cell locks open as soon as they were able to enter the area. By 8:00 p.m. the fire was officially under control.

It was, of course, believed by prison personnel that the fire had been deliberately set by prisoners to escape. Some of the later investigations questioned that belief, preferring instead to believe that a short circuit in electrical wiring caused the holocaust, but the actual cause was never proven. In the investigation that followed the tragedy, State Attorney General Gilbert Bettman demanded that Governor Myers Y. Cooper remove Warden Thomas. The prisoners, too, blamed the warden as somehow being responsible for the fire. The governor refused to

Convicts, guards and firemen work together to aid the injured prisoners removed from the death-trap cell blocks at the Ohio Penitentiary. The flash fire killed 320. Wide World.

Firemen try to tame a stubborn fire in a huge military records storagehouse near St. Louis on July 12, 1973. Several hundred thousand military personnel records were lost. United Press International.

remove the experienced warden and Thomas was quoted as saying: "If I was to blame for anything, I'd like to know it. I did three things: called the fire department, ordered release of the men from their cells and sought the protection of the public. Those things I had in mind when the fire began and those I attempted to carry out." The prisoners, however, testified that at least a half hour had elapsed between discovery of the fire and opening of the cells. They wanted the warden ousted. So heavy was the threat of more violence that more troops were called into the prison for protection.

The state had sentenced the men to punishment in the penitentiary. In doing this the state was responsible for their safety and certainly for the reliable construction and protection of their "home." Yet a fire drill was never held and fire procedures were never decided. And in the wreckage, overturned checker tables and half-eaten candy bars attested to

the swiftness of the disaster — 320 men had been unknowingly sentenced to death.

Other states have owned firetraps costly to human life. On April 13, 1918, thirty-eight people died at the Oklahoma State Hospital for the Insane at Norman, just south of Oklahoma City. The fire began at 3:00 a.m. from defective wiring in a linen closet, and flames quickly spread throughout the first floor of the two-story structure. Only a few patients, those near exits, escaped unharmed. Fire fighters were unable to control the fast-spreading fire.

On July 16, 1967, thirty-seven men were killed in a fire at a prison camp at Jay in northern Florida. The fire began during a brawl between two prisoners who, while fighting, broke a gas line. The gas ignited when a fluorescent lamp broke and eight minutes later the entire building was destroyed. And execution en masse was again provided by the state in lieu of precautionary measures.

Devastation at the Cocoanut Grove

A cheerless rain fell on the afternoon of November 28, 1942. It was the day of the big game.

The Boston fans were packed into Fenway Park to watch the hometown favorites, the football eleven of Boston College, triumph over their archrivals, the challengers of Holy Cross. Denny Myers, the Boston College coach, had outmaneuvered his opponents all year. The team was unofficially invited to the Sugar Bowl and elated Boston sportswriters were going to nominate the whole first string for All-American. The Sugar Bowl Committee was there to watch. Buck Jones, the Western hero who had come out of retirement to promote the sale of war bonds, watched the game from the mayor's box. The game was a sellout with an overflowing crowd of 42,000.

Plans for Boston College's victory celebration were arranged at one of Boston's most fashionable nightclubs, the Cocoanut Grove. The mayor would be there. Even fans who were not members of the night-club circuit made plans to be at the Cocoanut Grove that night.

But in the first ten minutes of play, Bobby Sullivan of Holy Cross dove across the goal line for a touchdown. Johnny Bezemes of Holy Cross dominated much of the game, once running 67 yards for a touchdown. Denny Myers's "T" formation couldn't get off the ground, even with wings. When the last quarter was over and the last cleat had churned the mud at Fenway Park, the score was 55-12 in favor of Holy Cross.

The mayor declined to go to the Cocoanut Grove, and Boston College canceled its plans for a victory celebration on the Cocoanut Grove Terrace.

But many members of the overflowing crowd who watched the unexpected happen between the goalposts that afternoon discovered that, after the initial disappointment, they were still in a mood to have a good time. And the gaity of the Cocoanut Grove

Firemen happened to be in the vicinity of Boston's Cocoanut Grove and responded before an alarm was turned in. Though dense smoke poured from the building (left), firemen's work centered around rescue. Military personnel and civilians aided in rescue efforts (opposite). Both Wide World.

certainly lent itself to the occasion. Although the day would end with grim finality, the evening began normally.

Cowboy star Buck Jones, who made bandits behave, had come down with the sniffles. Yet nighttime found him and his entourage relaxing in the Main Dining Room of the Grove. The trainer for the star-crossed Boston College team and his wife showed up, too, despite the fact the victory celebration had been canceled.

A group of eight women had saved all year for that night at the Cocoanut Grove, and they brought their husbands, who were devout Boston College fans, for their own victory party. Another group of five people who had watched the dismal defeat at Fenway Park found themselves at the nightclub, ending their day in the basement Melody Lounge. They were probably referred to as a nondescript "party of five" as the attempt was made to seat them somewhere in the densely crowded lounge.

A couple had hosted a party for football fans in their home after the game. The Cocoanut Grove, it was decided, was the best place for the party to end. Some members of the Cocoanut Grove band had been at the game also. There was a delay in the show and they had a chance to discuss the events. A young girl had been the guest of her brother and his wife at the football game. The three now found themselves seated in the splendor of the Main Dining Room.

The structure of the Cocoanut Grove itself was not likely to arouse the fears of anyone concerned with fire. It was composed of brick and stucco and was, in fact, "fireproof." Although the interior decorations were composed of materials long known by the human race to catch fire readily, eight days previously, Lieutenant Frank Linney of the Boston Fire Department had inspected the premises and described the conditions there as "good." He noted that there was "a sufficient number of exits" and a "sufficient number of extinguishers."

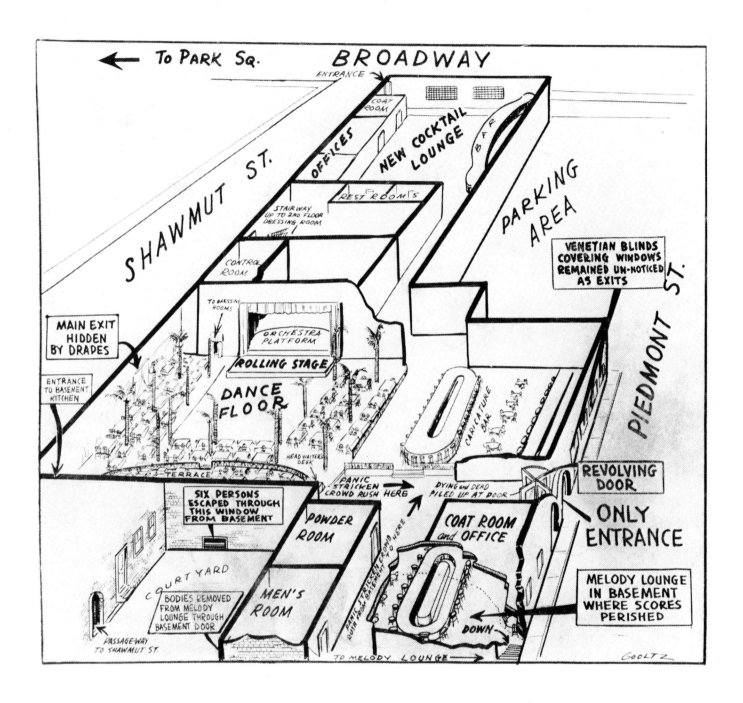

Anyway, the hazards of fire were not likely to be on the minds of the Cocoanut Grove patrons on the evening of November 28. Their primary concern was to have a good time. Enough worrying had been done at the football game that afternoon. Besides, there was a war on and for the men at the many bases near Boston — Fort Devens, Camp Edwards, Fort Banks and others — who knew what was ahead? There were many colleges around Boston — Harvard, M.I.T., Boston College and others — and there was just no finer place than the Cocoanut Grove to take a date.

So the Cocoanut Grove that evening was swollen with people. Its official capacity was between 400 and 600. That night waiters found themselves setting up extra tables on the Main Dining Room dance floor for people who would not be turned away. It was later estimated that the actual crowd was about 1,000.

Upon spinning through the revolving door, the main entrance to the building, this throng had entered a splendid world of palm trees and tranquil blue skies. The palm trees and other decorations were made of cloth, paper, rattan and bamboo, and the blue skies overhead were really blue satin. This paradise had walls — many were false walls — covered with imitation leather. Lieutenant Linney had touched a lit match to these same decorations just eight days previous. They did not burn.

The fire began in the Melody Lounge, where the only known exit was the stairway. Those who managed to make it up the stairway and through the narrow corridor crowded the revolving door, once the turnstyle to gaiety and carefree times for many. In addition, those from the Dining Room — Dance Floor area also jammed the revolving door or found their way to the exit in the New Cocktail Lounge. But here they also jammed up because the door opened inward instead of out. National Fire Protection Association.

Once through the door, patrons stepped into a foyer where they could check their coats. From here they could go through a narrow hallway and down a tight stairway into the popular Melody Lounge for a drink. They sat in the midst of palm trees that concealed electric lights in their fruit. The kitchen was also in the basement but it was so skillfully camouflaged that no one in Paradise would ever need to be confronted with the reality of a kitchen.

Or customers could go to a bar just off the foyer on the first floor. If they wanted food, drink and a floor show, they would go to the Main Dining Room. This room contained a raised Terrace from which patrons could watch the show in elegance and comfort. It had a dance floor immediately in front of the stage, and the roof above the dance floor could be opened in warmer weather to allow for the romance of dancing under the stars. Many customers were curious to see the new addition to this magnificent oasis called, simply, the New Cocktail Lounge.

The entrance to the New Cocktail Lounge was through a passageway at the far end of the Main Dining Room. There was also a door on Broadway Street to the New Cocktail Lounge, a door that opened inward. There were windows in the New Cocktail Lounge, but they were bricked with glass, possibly because the view from the windows might destroy the illusion of Tropicana.

There were no ominous clouds in the blue satin sky of the Cocoanut Grove in the early part of the evening. At 10 p.m. all was well. Shirley "Bunny" Leslie was selling cigarettes to the crowd in the Main Dining Room. Show biz was not unknown to her; she was the half sister of actress Lillian Roth. Stanley Tomaszewski, a waiter, was busy earning tips in the basement Melody Lounge. He was sixteen years old, and his presence there was illegal. But the Cocoanut Grove had experience in flirting with the law; its previous owner, a gangster, had been murdered.

Shortly after ten o'clock, Stanley was called upon to replace a light bulb in one of the graceful palm trees. He had difficulty seeing the socket, so he lit a match. As soon as the socket was located, the match was blown out. The flame did not come near the lovely green fronds of the tree.

The light bulb was replaced. Yet shortly afterward a small dark hole appeared on the ceiling. It widened and curled its edges in a black sneer. The previously tranquil palm tree was now being nipped by playful flames. Someone was dispatched with a bar rag to smother the dancing flickers.

The patrons watched the tiny fire. Though most were standing now, there was no move for the stairway, the one known exit.

Someone had apparently located one of the "sufficient number of extinguishers" and brought it to work on the now threatening fire. An attempt was made to fell the palm tree. When it was finally brought down, it was already too late. The cloth ceiling was already curling with flame. Someone allegedly grabbed a piece of the burning sky. When it burned to nothing in his hands, the serious nature of the situation became a reality. The sky was now raining sparks onto the heads of patrons. At first the voices had merely been charged with tense excitement. These crescendoed to screams. There was a cry of "Fire!" and a sudden attempt to reach the stairs.

Those who might not have thought of leaving just moments before now found themselves tangled in confused panic struggling to reach the exit. Many were crawling, many were attempting to run. The floor was unfamiliar terrain, and the Melody Lounge was now in choking darkness. It is not known whether this darkness was due to the lights suddenly going out or the thick, black envelope of smoke which now was smothering the terrified crowd. Most of the crowd in the basement attempted to flee through the one known stairway. Some found the kitchen door and escaped. Some survived inside the walk-in refrigerators. Some tore at windows in the false walls, only to find an impasse on the other side. Some were quickly overcome by carbon monoxide and what is believed to be some other mysterious gas. They were overcome so quickly that they were found later either standing at the bar or seated at their tables as if events had happened too quickly for them to even make a move to egress.

The mob clamored to the stairway. Women were fleeing with their dresses and hair afire. Men, too, were seriously and painfully burned. A door opening

directly to the outside was located at the top of the thirteen stairs. The first people to reach the top pushed on it but it would not open. They kicked it but it would not open. It was locked. Management policy: No one was to enter or leave Boston's finest night spot without paying.

The crowd, frantic now, stampeded from the top of the stairway toward the foyer and the main exit, the revolving door. An employee was already there trying to unfasten the revolving door's cable. It was no use. The crowd, now beyond any form of reason, quickly jammed the door. After the fire, two hundred bodies were found piled behind this exit.

The screams of the Melody Lounge crowd were not detected by the people upstairs in the Main Dining Room. Business went on as usual.

The flames consumed the surface of the basement walls and rode on the backs of the crowd trying to make their escape. The fire followed them to the foyer and invited itself into the Main Dining Room. The patrons in the dining room were waiting for the show to begin when they were greeted by a sudden rush of fire. The fire moved with such dedicated swiftness that survivors were later to remark that the air itself appeared to be on fire.

The dance floor, once the setting for gracious open air dancing, was now the scene of grotesque deterioration as people furiously attempted to escape. This area was also in darkness. Couples attempted to keep together, but the panic and the flailing arms, legs, hands and feet groping in the blackness, made this impossible. Most of the people in the Terrace section of the Main Dining Room did not escape. Here they later found the bodies of the cigarette girl and the Boston College trainer and his wife. Many, too, were quickly overcome by the gases bred in the churning cloud of dark smoke. Fortunately, some escaped through an exit to Shawmut Street.

Surprisingly, there were few injuries attributed to the panic itself. In a pamphlet put out by the National Fire Protection Association in 1943, Robert S. Moulton, the Association's Technical Secretary, notes that indications were "... people died too quickly to fight for their lives."

Many in the Main Dining Room had impulsively headed for the revolving door, where they became part of the most intense panic. It was here that many finally succumbed to the gases and a few were burned to death.

Black smoke billowed into the New Lounge. Many fled from the Main Dining Room into this area. Here, too, there was a locked exit that could have saved some lives. The surge was then to an exit on Broadway Street, the door that opened inward. Though the crowd pressed against the exit, a few still managed to escape through it.

A fire alarm box about a block and a half away from the Cocoanut Grove was pulled at 10:15, about the time the fire began in the nightclub. But it was not for the Cocoanut Grove: The alarm concerned a fire on the seat cushion in a car. Fire fighters quickly subdued this fire and, hearing the desperate screams from the direction of the Cocoanut Grove, they moved their apparatus there. The first alarm turned in specifically for the Cocoanut Grove occurred at 10:21.

A fire chief arriving on the scene and seeing the terrible destruction skipped a second alarm and pulled a third. A total of five alarms were sounded.

The work of the fire department centered around rescue activities. The fire fighters, however, were hampered in several respects. Just as the Cocoanut Grove was difficult to leave in an emergency, so was it difficult to enter. And the fire had to first be contained before it was safe for the fire fighters to enter. By 10:35, after the fourth alarm had been sounded, fire fighters were trying to master the fire at strategic locations. Entry into the New Lounge was attempted but bodies blocked the entrance. There were, perhaps, one hundred victims here. The glass brick windows were difficult to remove.

An upright bongo drum in front of the wreckage (opposite) serves as a reminder of what once was. The column supporting the roof was disguised as a palm tree; little remains now but the "bark." National Fire Protection Association. Sheet music and ladies' burned shoes (right) form an unforgettable epitaph for the 491 who lost their lives. Wide World.

151

Shattered glass, seared interior and grotesque wreckage on the sidewalk comprise a grim view for the curious. National Fire Protection Association.

Those found still alive inside were sent primarily to Boston City Hospital, where a victim arrived approximately every eleven seconds. They were given the best medical treatment available at the time, yet many placed in white hospital beds died, some in great pain. Much was learned from this ordeal about the treatment of burn victims.

Buck Jones died at Massachusetts General Hospital. Of the group of eight women and their husbands that had waited for this evening, only four escaped alive, and one of those later died. The girl who had gone to the game with her brother and sister-in-law was dead. Her brother lived, but his wife had died.

The fire had moved with unsurpassed rapidity. It took only twelve minutes for it to do its damage, and 491 people died as a result. The space for rescue was limited at the scene. Bodies were stacked on one another and placed in a nearby garage.

The cause of the fire was not known for certain. It could have been started by Stanley's match, or it could have started by a careless cigarette or match anywhere. Later testimony in the subsequent investigation revealed that the Cocoanut Grove's electrical wiring was installed by an unlicensed electrician. So perhaps the fire was started by defective wiring. Another theory was that the great amounts of alcohol

Firemen search the ruins of the New Cocktail Lounge. The Main Dining Room is down the hallway in center. Damage was complete and the building was later razed. Wide World.

consumed charged the air with vapor which spread the fire rapidly, but this is unlikely. The building was once occupied by a motion picture film exchange, and it was argued that the mysterious gas could have been produced by the ignition of film. There were numerous other theories, but officially, the flash fire that killed nearly five hundred people was listed as being of "undetermined origin."

How the fire started was not as important as why it had such tragic consequences. The public demanded answers, and several investigations were held. The grand jury indicted ten men to be held responsible, but eight out of the ten were acquitted. The club owner, Barnett Welansky, was sentenced to twelve to fifteen years. The contractor who built the New Cocktail Lounge was sentenced to two years on a charge of conspiracy to violate building laws.

Throughout the country laws were changed, building codes updated. Yet a member of the National Fire Protection Association said, "The majority of building codes throughout the country are antiquated and incomplete."

It was the day of the big game. Boston College was expected to win, but they lost; the unexpected happened. That night at the Cocoanut Grove the unexpected also happened, and 491 people lost. But they lost much more than a football game — they lost their lives.

Fire Under the Big Top

On the hot, humid afternoon of July 6, 1944, in Hartford, Connecticut, approximately seven thousand people, mostly children, crowded into the big top of the Ringling Brothers, Barnum & Bailey Circus for a special matinee. The previous day's performance had been cancelled because the circus arrived six hours late from Providence, Rhode Island. In order not to disappoint anyone wanting to see the "greatest show on earth," another performance had been scheduled. The main entrance was at the west end of the nineteen-ton tent and the people flocked through it by the thousands amidst the smell of straw and wild animals and the sound of children's voices pleading for pink lemonade.

Most sat in the bleachers while those with reserved seats sat on folding chairs in front of the bleachers. On the long north side of the tent were three outlets, but these were blocked by chutes used to escort the wild animals from their cages to the tent. On the south side were three exits, one of which was blocked by cables.

The circus began its history in America in 1790. Since then, more than one thousand circuses have toured the country. Circuses have been a colorful part of America that few have been able to resist. Even George Washington attended circus performances in 1793 and 1797 and sold a horse to John Bill Ricketts, the man who introduced the circus to the country. In its history, the circus had been plagued with numerous disasters including fires, wrecks and storms. In the nineteenth century P. T. Barnum's Circus, Museum and Menagerie burned so frequently that it was nearly impossible for the organization to obtain fire insurance. Other circuses and circus camps were victimized by fire but always the tradition of "the show must go on" prevailed.

Yet tragedies do occur, and, just in case, the huge circus at Hartford had taken certain preventative steps. The circus owned many buckets and portable fire extinguishers. The fire extinguishers, however, had not been distributed throughout the tent on the hot afternoon of July 6. There were four trucks on

(continued on page 157)

The big top was positioned on center support poles and could hold between nine and twelve thousand people (top). With the smell of sawdust and candy in the air, people packed the bleachers (above) and were often unable to decide which ring to watch.

During a matinee performance at Hartford, Connecticut, on July 6, 1944, a flame playfully tripped up the tent's side and in a few minutes the tent was a roaring inferno. The terrified people who managed to escape ran to safety just before the big top collapsed. United Press International.

hand: Three had a water capacity of one thousand gallons and one had an eight-hundred-gallon capacity.

There were certain unfavorable factors. Circus personnel took good care of the tent; it had cost $60,000 to purchase. The canvas for the big top had been weatherproofed the previous April with a coating of paraffin which had been thinned with gasoline. For some reason, the Hartford fire marshal had neglected to inspect the tent before the matinee.

The performance began. The second act in the show was Alfred Court's wild animal act which the crowd enjoyed immensely. As the animals were being escorted to their steel-mesh enclosures that led back to their cages, the Flying Wallendas, the famed balancing act best known for its accomplishment of the seven-person pyramid on the high wire, were climbing the tall poles and readying themselves for their act. Emmett Kelly, the sad-faced hobo clown, was busy going through his antics making the children and their parents laugh. He was one of the stars of the circus and a favorite of the people. But he never smiled during a performance, preferring instead to make others laugh with his deadpan expression.

A spot of flame appeared on the tent at the main entrance. A Hartford policeman on duty there said that when he saw it, the hole it made was no bigger than a cigarette burn. Slowly the tiny flame tripped up the thin canvas wall. It increased in size and traveled toward the tent's roof. It was still a small fire, however, and most of the audience and performers were not yet aware of it. Spotlights briefly focused on the Wallendas.

Merle Evans, the circus bandleader, saw the fire at about the same time the patrolman saw it. He instantly led the band into the lively "Stars and Stripes Forever." This was the traditional call of alarm in circus language. Circus personnel immediately went to their posts. Someone threw three buckets of water onto the fire, to no avail. Trainers tried desperately to hurry the wild animals out of the ring. The beasts were being encouraged by all means available to their demanding trainers to enter the gangways. One trainer turned a hose on leopards reluctant to leave the ring. All personnel knew that any impending tragedy would be made worse by wild animals in the tent's center. The Wallendas descended speedily on their ropes and tumbled to the ground to safety.

The crowds who previously did not know that the band was playing a "disaster march" now were undecided whether to watch the trainers struggling to remove the animals or to watch the increasing fire. Buckets were still being thrown on the blaze which

P. T. Barnum's Menagerie burned often at the turn of the century, but, unlike the Hartford fire, loss of life was restricted to the animals.

had climbed to a height of five or six feet. Circus hands were animated, trying to decide what to do. Perhaps that is why there was no immediate panic by the audience: It was a small fire and it was being dealt with by people who surely knew what they were doing. Though the fire was growing and was about two feet in width, there was still no migration to the exits. The wind, however, suddenly picked up and the fire began to edge across the top of the tent with alarming rapidity.

Support ropes were burned and the tent's six huge support poles began to fall, tearing flaming pieces of the canvas with them as they toppled, injuring and burning many. Scores managed to escape through the main entrance before it was completely engulfed in flames. Hundreds of parents dropped their children from the back of the bleachers and then jumped themselves. Children, separated from their parents by the confusion, wailed and screamed. One little boy hovered over his fallen grandmother, pleading with the stampeding crowd to help him lift her to her feet. Patches of flaming canvas continued to fall and women, their hair and clothing on fire, shrieked and moaned. Those seated in the folding chairs tossed their seats aside, causing others to trip over them, forming a human barricade to prevent any escape. There was a rush to the three exits on the north side. Hope was lost to those choosing this route as they discovered the gangways were for the exclusive use of lions, leopards and tigers, many of which were still in the chutes on their way back to their cages. Hundreds of bodies were later found here. One by one the heavy support poles crashed into the crowd. The sixth one fell, and the entire tent, wrapped in flames swooped onto the crowd, blanketing them in the burning canvas. Those screaming beneath the fallen tent were doomed and soon they were silent. It had taken the flames only ten minutes from the time the first cry of fire was shouted to inflict their damage.

The fire department was summoned to the scene and, though firemen arrived almost instantly, they were too late to save lives. All they could do was spray water on charred ruins. But fire fighters found that there were no hydrants on the circus grounds; they had to couple their hoses with hydrants outside the area, three hundred yards away.

Ambulances lined up to take victims to the hospital. Hartford hospitals were prepared for such a disaster, since World War II was in progress and major hospitals were instructed in burn treatment in preparation for air raid victims. The same type of treatment had been used successfully two years previously in the Boston Cocoanut Grove tragedy: Victims were
(continued on page 162)

Sad-faced Emmett Kelly, a master clown and long a crowd favorite, carried buckets of water after the fire to help snuff out the still-burning embers.

159

This aerial view of the big top site shows the disfigured bleachers, broken arena rings and charred center poles. Wide World.

Bros AND BARNUM & Bailey

IRCUS

BLE ALFRED COURT MASTER TRAINER

Alfred Court had just finished his wild animal act and the beasts were being escorted to the chutes that led to their cages when the fire started. People divided their attention between the animals and the Flying Wallendas who were climbing to the high wire, although some people had turned to watch the growing circle of flame. The tragedy could have been worse if the animals had been still performing.

given morphine, wrapped in sheets and given plasma injections.

The dead numbered 163 — half of them children. All of the victims were spectators who had come to the big top for an afternoon of carefree fun. All of the circus people escaped, though the Wallendas just barely freed themselves. Emmett Kelly, his sad face looking even sadder, was seen after the fire carrying buckets of water for the bucket brigade that was putting out sparks and embers.

The rapidity with which the fire moved was blamed on the improper weatherproofing of the canvas. After a brief inquiry, five Ringling officials were arrested and warrants were issued for four more. Later, seven of the defendants received one-year prison sentences.

Days after, flags flew at half-mast throughout the Connecticut River Valley. Some Hartford funeral parlors held services at fifteen-minute intervals. The burning of the big top on July 6, 1944, was the worst in circus history, one that people did not soon forget.

Bleachers that once had been filled with children who had looked forward for days to seeing the wild tigers and other circus acts in the undisputed "greatest show on earth" (left) were burned and broken (above) in a mood of tragic desolation. Wide World.

Gushers of Flame

Before 1859, oil was little used because of its scarcity. It had to be soaked from streams with blankets or pumped from creeks, and the idea that rich oil beds lay deep beneath the earth's surface had not occurred to anyone. A salt company discovered underground oil by accident, and their ad in a New York paper advertised petroleum as a "wonder drug" since it was extracted from salt beds four hundred feet below the earth's surface.

The first real quest for oil began when Colonel E. L. Drake dug an artesian-type well at Titusville, Pennsylvania, on May 20, 1859, ten years after many Pennsylvanians left to seek gold in California. On August 27 Drake's well, the first in America, began to flow oil, and the Pennsylvania oil rush was on. Though the well's depth was only sixty-nine-and-a-half feet, there was an abundance of oil in the well — up to twenty barrels were pumped every day.

Oil was first used by ancient Greeks and Romans to feed the sacred temple fires. Prior to the discovery of oil in Pennsylvania, however, whale oil was used to light lamps in America. But the whales became scarce and a substitute had to be found. Oil seemed to be the answer and in 1859 two thousand barrels of oil were produced. Just one year later production had increased to five hundred thousand, and by 1874 the annual yield was six-and-a-half million barrels.

As westward expansion increased, oil was discovered in Kansas, Oklahoma and Texas. Small Western towns frequently voted for bonds to build a city hall and used the money to dig for oil. Sometimes the idea made the town wealthy and sometimes the money ran out too soon. Rags-to-riches stories flourished — one school district managed to reap an income of $22,000 a year upon discovering oil in the school yard. A poor farmer, attracted by the lure of the West, moved to Kansas with only a few possessions; oil was found on his Kansas land and he soon drew a one-thousand-dollar-per-day income. Many people just stood aside and waited for the drillers to usher in their well, while many others, after waiting too long, admitted defeat

When a gas pocket is hit during the drilling operation a huge and costly gas-well fire results. This 1964 New Mexico fire was fed by sixty million cubic feet of natural gas each day. Wide World.

in America's greatest gamble. But the Western landscape soon became dotted with oil derricks, and the noise of the drills or pumping engines filled the air.

Drake's well went up in flames six weeks after it was dug. Hastily scooped dirt put out the fire in his well, but as wells were dug to depths of ten thousand feet and more, oil fires got bigger and the search for effective fire-fighting methods began.

The fires themselves are terrifying. Usually they start from lightning, friction in the machinery or spontaneous combustion from old rags lying in the area. Before improved fire prevention methods, an average of one fire a month was due to lightning.

Occasionally when digging a well, a gas pocket is hit and the well, instead of blowing "in" with a torrential gusher, blows "up." Escaping gas pours from its imprisonment beneath the earth with a

Before America became conscious of preserving its natural resources, oil well fires sent great clouds of smoke into the air for days as unconcerned onlookers posed in the foreground. This fire took place at the turn of the century near Houston when Texans were proud their oil fields were so rich that there was "oil to burn." The company that owned this well claimed it refined the purest oil in use. University of Texas.

deafening roar, vibrating the ground. The noise is so loud that it can be heard from a distance of twenty miles. Fire fighters, often the drilling crew who don fire-fighting attire, are sometimes deaf for a week after a big fire is extinguished. Professional oil well fire fighters, who often work at the tough blazes in thirty-six- to forty-hour stretches without a rest, usually become hard of hearing after a few years.

While there is no oxygen for combustion within the well itself, the gas can be easily ignited at the surface by as simple a cause as a stone striking the casing, creating a spark. Some fires touched off in this seemingly harmless way have lasted more than three years and bankrupted many oil companies. The light from these fires is so bright that observers claim that you can read a newspaper by their light as far as a mile away.

When the well explodes, the platform is often immediately enveloped in one- to three-hundred-foot-high flames with a seemingly endless fuel supply provided by nature deep beneath the earth's surface. If the crew that was on the equipment platform is fortunate enough to be still alive, they dash from the area until they are safely a quarter mile away. Many times they must run farther because debris from the well equipment and rocks lifted by the tremendous pressure begin to crash around them. The derrick becomes white-hot and topples to the ground, seemingly without a sound because the noise of the escaping gas is too loud. Sometimes a fire does not immediately appear after a gas pocket is hit, but the danger of one starting is always there.

In the early part of this century, fearless Tex Thornton was at the scene of many fires. He was the

In the days before oil-well fire specialists, steam and sand were used to extinguish the difficult well fires. The Spindle Top oil well district in Texas near the Louisiana border had many serious fires. This photo was taken in September 1902 when a well burned for a week. University of Texas.

daring fire fighter of the oil industry and one grateful oil commissioner commented in 1928: "In the past three or four years, just to give you some idea of his value to the industry, he has saved enough natural gas alone, not to mention oil, by fighting fires no one else would tackle, to run the city of New York more than a hundred years." Tex, who boasted some four hundred scars and a permanent tan from the intense heat that accompanied his job, wore an asbestos suit and sometimes a gas mask when fighting a blaze. The fireman was accustomed to performing before large crowds, rarely working before less than a crowd of a thousand and often before a crowd of ten thousand.

He seemed unbothered by the fact that he was unable to obtain life insurance, and continued to dart directly into oil well fires, first placing cables on the twisted, hot masses of metal which had once been part of the well's rigging, then hauling them away since the heat from the hot metal could rekindle the fire once it was out. A man who lived in the mouth of danger, he could console himself by his adequate compensation for his task. "An oil well fire is an expensive proposition," he explained. "That's why I occasionally get paid up into five figures for puttin' one out." He was known by his admirers as the "one man fire department of the oil fields" and, for the most part, he accomplished his tasks singlehandedly.

A team of two brothers was just beginning a career of putting out fires in oil wells in the early 1920's. Myron and Floyd Kinley were introduced to the excitement of the oil fields by their father, Karl, who was an oil well "shooter" who used dynamite to make, or "bring in," new wells. He was good at his job and was called to Taft, California, in 1913 to try to cave in a burning well with the use of explosives. The fire was extinguished and the technique of exploding wells to snuff out the fires came into use.

Oil well fires stream their furious flames out of a pipe that measures only a few inches in diameter, contrasted with land conflagrations that can envelop an entire city. Yet oil blazes are some of the most difficult to fight since water is practically useless on them. All fires need oxygen to burn and one of the best ways to put out a fire is to eliminate the oxygen supply. The use of explosives in well fires displaces the oxygen supply mixing with the gas in the stubborn fire column. Once the oxygen is removed, the fire cannot burn.

Wells usually do not strike underground pools of oil; rather they strike sand that soaks the oil thousands of feet beneath the surface. In earlier days of well drilling, the opening at the drill's tip had to be enlarged to enable the crew to pump the oil. In the 1920's Tex Thornton "shot" the wells with nitroglycerine to make the opening larger. And if the drilling operation caught fire, Tex was also on hand to put the fire out. After hauling away as much burning debris as possible, he would proceed to fight fire with fire as he took nitroglycerine shells from three to ten feet long and holding ten to thirty quarts of the chemical in each to a point eight to ten feet above the well's mouth. The explosives were set off and the fire, if Tex was lucky, was out. Nitroglycerine, one of the most powerful explosives known, was used by the fire fighter in dosages of up to 1,200 quarts. Once the charge was set off, there was nothing he could do to prevent whatever were the consequences. "No use running from that stuff," Tex once drawled. "If it's going to go off and you are near it, it'll get you, no matter how fast you run."

One time Tex placed 150 quarts of nitroglycerine on a well's platform in preparation to bring in a new well. He decided to rest awhile and left the scene briefly, going to an engine house about sixty feet away. A few minutes later a nitroglycerine shell blew into the top of the derrick and exploded with a terrific blast. Iron and timber rained over the area, showering the ground with debris. Tex rushed outside and discovered that the ground was covered with fiercely burning oil, and seven round objects were lying on the ground in the burning inferno. After quickly soaking his clothes with water, he ran through the flames. The seven "objects" turned out to be the crew, and it was too late to save them.

The Kinley brothers bore the scars of many narrow escapes, and still continued in their profession. In 1924 two wells caught fire simultaneously in Cromwell, Oklahoma. An experienced crew put out the first fire, but the second called for the valor and fire-fighting knowledge of Floyd and Myron Kinley. After eighteen hours of hard work, they finally snuffed out the flames.

In October of 1929, just outside of Oklahoma City, smoke blacked out the sun during the day and red-orange flames lit the area at night. An oil company's well had struck gas unexpectedly and the ensuing fire threatened to devastate the rich Oklahoma oil field. The heat had melted the derrick into distorted shapes of steel. A religious sect devoutly prayed, fervently believing it was the end of the world. Called to the scene, the Kinley brothers brought thirty quarts of dynamite and proceeded to place it above the heart of the fire through the use of a makeshift boom, protecting themselves with only an asbestos-lined shield, since experience had shown them that asbestos suits were too bulky. The brothers scurried away from the blaze and detonated the blast. After an earth-shaking roar, all was silent — the fire was out.

In 1931 a wild well fire in the East Texas oil fields near Tyler defiantly burned for two hundred hours and took a toll of nine lives. The well was a gusher and its uncontrollable flow spewed from beneath the earth. As men tried to subdue it, the gusher suddenly turned into a flaming torch. The crew was knocked down by the force and some were immediately overtaken by flames. An eyewitness, who later died from his burns, said: "I was standing about halfway between the edge of the derrick and the hole. Suddenly the elevator fell. The burst of flame was immediate. The falling elevator must have caused a spark that set fire to the oil and gas fumes. I started to run. I found my clothes on fire. I rolled on the ground to put out the fire, but that did no good."

Some of the men who managed to get up from the derrick floor ran through the woods and called for help. Bystanders grabbed the fleeing victims and tore the flaming clothes from their bodies. Again asked for help, the Kinley brothers finally snuffed out the fire with two hundred pounds of nitroglycerine.

Several times in their careers the brothers had close calls with death but somehow always managed to escape. At a big oil well fire near Goliad, Texas, in 1937, Floyd Kinley did not escape the clutch of death in time and died in the blaze. Myron still fought the risky fires, but refused to allow his stepsons, who used to assist him during school vacation, to work with him anymore. Kinley did not want anyone in the family — besides himself — to risk his life at the hazardous fires. He went on to become the highest paid fireman in the world, and a millionaire. He feverishly worked at about five hundred of the big ones, fires that burned millions of dollars worth of equipment and irreplaceable raw materials, and he fought them on every continent in the world. His job confined itself strictly to oil and gas well fires; he was the first to admit his ignorance in fighting land fires.

Kinley was greatly appreciated by the oil companies. A big oil well blaze can do billions of dollars damage, making it the costliest of all industrial disasters. One oil company executive called him the "one indispensable man in the entire industry."

In the 1930's a well named Wild Greta raged out of control for three and a half years in Refugio County, Texas. The well had "cratered" or caved in a hole three hundred feet wide and eighty feet deep. Myron Kinley was successful in putting out the fire and sealing the well in 1936, but the task took him eight months to accomplish.

A friction spark from the drilling rig after the drill hit a pocket of gas at a depth of nearly ten thousand feet was the apparent cause of a fire at Elk City, Oklahoma, in 1950. Its bright flames could be seen for fifty miles, and the petroleum industry knew that Kinley was the only person capable of snuffing out the fire. "I get a knot in the pit of my belly every time I go into a tight spot," the scarred fire fighter

once said. Kinley used his explosives in this fire, but upon detonation of the charge, the fire was only temporarily extinguished and started again four minutes later. As he prepared to set off another charge, the sand in the well collapsed and choked the fire.

Later that same year an oil well, Holley No. 1 near Big Spring, Texas, caught fire from a spark of a car's exhaust. Holley became a geyser of flame and it took Kinley and his crew twenty-five days to put out the stubborn blaze. Nearly half a million dollars went up in flames and it cost the oil company nearly that much to put it out.

When an oil well caught fire off the Louisiana coast in 1953, Kinley again was on the scene. Upon studying the situation, he decided that the "Christmas tree," the series of valves attached to the casing that controlled the gas flow and caused the fire to flame out horizontally rather than vertically, had to be knocked off so that the fire would rise upward, allowing Kinley to put his explosives below. If the valve was removed, the fire could be confined to one jet, making fire-fighting operations much simpler. But the problem was how to displace the "Christmas tree." He borrowed an army rifle team from a nearby military base which sprayed the valves with bullets from morning until evening without success. Finally Kinley was hoisted by means of a water-cooled boom into the fire to retrieve the valve. He managed to snap it off but by this time the valve of another well in the area had cracked due to the high heat, and its escaping gas had also caught fire. After ten days of laboring he finally got the whole fire out.

Kinley continued to work long past retirement age. Once he jokingly said that he worked to provide for

gas. The oil and gas burst into wild flames and men began to don life jackets. The crew foreman grabbed the radio microphone and blurted, "Mayday! Mayday!" — the traditional call for aid. As the crew members looked into the water below, it appeared to be a sea of fire. Flames on the platform were attaining heights of three hundred feet, and soon the entire platform would be a mass of flames. There was no other choice but to jump. A few minutes later the coast guard and private helicopters had rescued the survivors in the water; four men had died.

Oil company officials did not want to extinguish the fire, since the blaze was protecting the coast from a disastrous, polluting oil spill. They decided to eliminate the fire at its source by cutting off the oil and gas supply twelve- to fourteen-thousand feet below the surface. Barges sprayed thirteen thousand gallons of water per minute on the platform as the decision was made to drill other wells in the area to the gas pocket and then shoot mud down the well. The mud would be sucked up into the flaming well, thus cutting its gas supply.

Company engineers flew over the area in the days that followed as fire-fighting efforts continued, looking for signs of oil spillage. Several plans were designed to lap up any leakage. Should all other efforts fail, twenty-two thousand bales of straw were arranged along the coast to absorb the oil.

The fire, meanwhile, continued to rage and spread to other wells whose safety valves had failed; soon eleven wells were burning. The plan for extinguishing the fire worked, however, and by April 16, 1971, the last flaming well was capped. The explosion had cost the company a five-million-dollar platform and twenty-nine million dollars in pollution control. And it had taken 136 days of work to put the fire out.

Lavoisier, the famous French chemist whose discoveries in combustion were a basis for modern science, said in 1786 that "fire is the combination of a substance with oxygen." Safety devices and better trained crews assure less fires in oil wells; very few new wells have been known to catch fire. Yet when oil is the "substance" and combustion takes place, the result is an explosive blaze of the first magnitude.

his family, since it was impossible for him to obtain life insurance. Those who knew him knew that he worked for the excitement and the satisfaction of knowing that only he could tame the untamable fires.

By the end of the 1950's, oil well fires were becoming increasingly rare. When they did occur they were fought with scientific and technical knowledge. Platform B was located sixty-five miles south of New Orleans in the Gulf of Mexico and was home to sixty men. The U.S. Geological Survey inspected the property and found everything concerning safety standards to be in order. It was the biggest offshore drilling operation in the Gulf and was considered to be the best in engineering design. On November 30, 1970, a drilling team was attempting to clear Well B 21 of mud. At 9:30 the following morning a loud explosion was heard and then the roar of escaping

Hospitals: Healing and Dying

In the dedication speech at the Cleveland Clinic in 1921, Dr. George W. Crile said that the purpose of the clinic was "to give assistance in solving the problems of the patient of today and through its investigations, its statistical records and laboratories to seek new light on the problems of aiding the patient of tomorrow." The founders of the clinic — Dr. Crile, Dr. William E. Lower, Dr. Frank E. Bunks and Dr. John Phillips — gave a $100,000 endowment and provided that one-fourth of their annual incomes would go to the clinic. Doctors Crile, Lower and Bunks had practiced medicine together for thirty-five years. Dr. Crile was a brilliant surgeon who was not lacking in idealism. He had been a doctor on the battlefields of France in World War I, and later went on to perfect thyroid surgery, a development widely heralded by the medical profession.

The four-story clinic became a proud Cleveland institution. On May 15, 1929, about three hundred people were within its walls. Some lay on operating tables, some rested in bed and some sat nervously in waiting rooms. At 11:30 a.m. a resounding explosion occurred in the basement where the clinic's X-ray films were stored, and the films immediately burst into flame. Several theories attempted to explain the explosion. A leaky steam pipe, authorities later reasoned, overheated and the highly combustible X-ray films in the same room caught fire. The room was equipped with a fire door but it failed to function. Other experts insisted that that theory was

The lawn that once surrounded the Cleveland Clinic with refined greenery was soon crowded with fire and gas victims removed from the building. Cleveland Press.

Throngs watched as firemen performed rescue operations from the roof of the Cleveland Clinic on May 15, 1929. The seething gas from burning X-ray films in the basement quickly claimed victims. Cleveland Press.

no more plausible than a carelessly discarded match or cigarette. No one was blamed; the impact lay in the tragedy itself.

Poisonous yellow gas fumes emanating from the burning film swirled in a fast-moving cloud throughout the clinic. People were quickly overtaken and gasped for breath, running to the windows to seek oxygen. The fire continued to burn up the air supply and, coupled with the choking gas, began to claim its victims. The fumes poured in through ventilator shafts, up stairways, through halls, and the fire found fuel in stairway woodwork. Windows burst and passersby on the street in front of the clinic were overcome by the fumes.

Witnesses on the scene after the explosion said that they could hear the terrified screams for blocks.

Firemen arrived immediately and attempted to enter the building but the gas fumes repeatedly drove them back. A battalion chief directed his firemen to scale the roof and enter through the skylight. Shortly afterward, two firemen lowered themselves from the roof into the building. It was not an easy task to enter. From the skylight, the firemen suspended themselves and then swung their bodies to gain momentum in order to drop with a minimum of injury inside the mezzanine rail encircling the fourth and top floor.

The two firemen, upon entering, found bodies packed in the space between the elevator and stairway where occupants of the clinic had tried in vain to escape. Other fire fighters working to save lives from the roof succeeded in opening the trap door there only to find a mass of bodies of people who had attempted to find refuge on the roof. One fireman was horrified as he looked into the building through the roof's skylight. "I hope never to have to look at anything so horrifying again," he said, obviously shaken. "Lord help me, as far down the stairway as you could see were bodies, bodies, bodies. Twisted arms and legs, screaming men and women. Bodies and screams."

The firemen managed to lift out fifteen survivors of the catastrophe, but the jam at that escape route was so great that many at the bottom of the pile were crushed to death. Pulmotors, instruments that contain oxygen to revive people, were quickly ordered to the roof. One battalion chief lowered himself into the building and was appalled at the condition of the people inside, some still barely alive. He ordered the firemen to concentrate nearly all their efforts on getting the trapped and overcome people out. The screaming was the worst on the third floor, one fireman on the scene reported, and firemen wanted desperately to reach the trapped people in order to save them. Pulmotors were brought inside and firemen with hose lines literally cut through the cloud of smoke and could be seen fighting the flames to protect their fellow firemen who worked to resuscitate overcome victims.

Many collapsed at the windows seeking their last breath of air. Both entrances to the street were blocked by the panic-stricken clinic patients and personnel. The fire did its damage quickly: Blistered plaster and a secretary's half-finished letter were found in an office, while in other areas surgical equipment was ready for use.

The clinic's yard, which surrounded a building whose mission was the saving of lives, soon became covered with the dead and the dying. Any available vehicle in the area was commandeered by the authorities to transport the injured from the burned-out hospital to other Cleveland hospitals. It took three hours to lift the bodies one by one through the skylight.

The gas from the films did not claim all its victims immediately. Some people walked out of the building healthy and aided fire fighters in rescue work, only to collapse and die hours or even days later. Dr. Crile, his hospital's front lawn reminiscent of a World War I battlefield, helped in first aid and rescue, and later visited the fire victims at city hospitals. His close colleague, Dr. John Phillips, one of the clinic's founders, was in critical condition. Dr. Crile gave his own blood to save his friend, but the doctor died despite all efforts to save him.

A professional football player helped with rescue operations on the scene, felt well and considered himself fortunate when he returned to his home. He died from the effects of the portentous gas forty-eight hours later. Several firemen were hospitalized. The clinic which had dedicated itself to patient welfare just eight years before was witness to 125 deaths that May 15.

Yet Dr. Crile responded to the tragedy with the traditional American optimism. His own personal ideals and belief in his profession allowed him to resume operation of his clinic within days in the temporary quarters of an old school due to the help of several influential Cleveland citizens.

Another hospital fire took place in a south-central Illinois community of eight thousand on April 5, 1949. St. Anthony's Hospital was located in the town of Effingham, and its 125 beds were adequate for the area it served, though most of the building was over sixty years old. A fire in a laundry chute was discovered just after midnight that day.

Shortly after the alarm was turned in, flames could be seen lashing their fiery tongues through the front entrance. People living in the surrounding neighborhood rushed to the scene and helped twelve to fifteen persons to safety. Soon hundreds crowded in front of the hospital but due to falling bricks and rubble were unable to get close enough to the building to perform

Effingham, Illinois, citizens watched numbly as St. Anthony's Hospital, an institution devoted to healing, burned in 1949. The catastrophe took seventy-seven lives. Wide World.

The mental health wing of Mercy Hospital in Davenport, Iowa, housed about seventy patients when it was gutted by flames on January 7, 1950 (above). Only the walls remained (below) as ruins smoldered. The fire trapped many and the death toll reached forty-one. Both Wide World.

The roof of St. Joseph's Hospital in Phoenix, Arizona, was vividly ablaze on October 5, 1917. The Phoenix Union High School football team averted a major tragedy by evacuating all the patients in time. Other people can be seen quickly removing some of the building's contents. Arizona Photographic Associates.

rescue operations. Screams of trapped patients, many of them friends and relatives of the people in the throngs outside, were painfully audible.

The second floor nursery was the first to be engulfed in flames. A nun who could not leave the twelve babies stayed and perished with them. Some patients jumped from their windows while other locked in casts and traction simply could not move and died when the flames reached them. There were thirty patients on the third floor and a nurse said that she believed none got out alive.

Fire fighters from Effingham and nine nearby towns helped to fight the fire, but low pressure in water mains meant a fifteen-minute delay in getting water on the fast-spreading blaze. Firemen pleaded

with the people not to jump, but as least a dozen did. Floors collapsed during the fire, and victims and equipment plunged downward in a tragic heap. Several explosions shook the area as ether tanks burst. The operating room was the only area not destroyed; only partial walls remained in other sections. One person commented: "Our fire fighters tried but the fire was so fast that I think it just burned itself out. It couldn't be fought." The hospital, popularly considered to be "fireproof," was destroyed within an hour after the first alarm.

A statue of St. Anthony holding a baby in his arms within the hospital was untouched by flames. The victims within the walls of an institution devoted to healing were not as fortunate — seventy-seven died. One citizen said, "No one in town hasn't lost a friend."

Burn, Baby, Burn

"How is it that we hear the loudest yelps for liberty among the drivers of negroes?" queried Samuel Johnson in the eighteenth century. The colonial British governors were encouraged to cultivate "the African trade" as the slave trade was modestly called. But great care was taken that the slaves were "properly watched, lest they should commit the odious and ungrateful crime of seeking to emancipate themselves by violence." The oppressors of the black population were continually imagining outbreaks toward freedom and throughout American slave history they believed that their own liberty was being threatened by the very people they gripped so tightly.

The city of New York cultivated the slave industry so successfully that the trade grew and flourished. Negro slaves were ill-treated at best and, it was noted, "caused their masters a great deal of anxiety." The masters, of course, were acutely aware of the treatment they were inflicting and consequently expected some widespread revolt.

New York in 1741 was a city of no sidewalks and few paved streets with an industrious population of ten thousand — one-fifth of which were slaves. The city had a fort that contained the governor's and secretaries' houses, with a British flag proudly waving above. Laws had been passed to prevent Negroes from assembling, and the slaves were often whipped and tortured. They were described in later accounts as being "constant objects of suspicion and fear." White slave owners came to fear slave contact not only with other slaves but also with the crews of ships that docked in the harbor. While the slaves were considered "barbarous and brutal," the ships' crews were considered to be "reckless and depraved" — certainly not a healthy influence on the "wild savages from Africa," as the slaves were often thought to be.

In March 1741, a Spanish ship with a crew comprised partly of Negroes was in the New York harbor. The Negro sailors were confiscated by the enterprising citizens of New York and sold at auction as slaves. The crew members understandably resented this since they had signed on board the foreign vessel as sailors. Their insolence was considered inappropriate and they were flogged.

On March 18, the governor's house in the fort caught fire. The day was "wild and blustering" and the wind bellowed the flames to the King's chapel, the secretaries' houses, and the barracks and stables. In addition, a house belonging to the master of a Negro sailor, purchased only a few hours before, was on fire. New Yorkers began to be suspicious.

A few days later, Captain Warren's house near the fort was on fire, and a few days after that, Van Zandt's storehouse burned. Then, three days later, a cow stall in John Murray's stable on Broadway burned, and within hours Mr. Thompson's house caught fire. A phenomenon of "universal panic" now gripped New York — "There seemed little doubt that they [the fires] had been the result of some secret plot." Cries of "It is the Spanish negroes! — Take up the Spanish negroes!" filled the air along with growing rumors. The Spanish Negroes were immediately "taken up" and thrown in jail. A fire occurred that afternoon and gave impetus to the idea that *all* the slaves were plotting to burn down the city.

One New Yorker came forward and mentioned that she heard Negroes wickedly threatening to start fires. Officials could not believe that the Negroes were the masterminds and they ordered the military to seek any concealed enemies that might be lurking about, encouraging radical behavior. Sentries were posted on streets and on April 13, houses and suspicious persons were searched. Since no concealed enemies were found, it was, the citizens thought, the Negroes who hatched this diabolical plot. Negroes were arrested, though all denied knowledge of any conspiracy.

A reward of one hundred pounds was offered for information concerning the plot. It was then that Mary Burton, an indentured servant, came forth with her story. She first implicated her master, John Hughson, a man of "infamous character." She noted that the slaves met at Hughson's tavern frequently. The topic of discussion was the burning of the fort,

Determined fire fighters in a snorkel train a healthy stream on an incendiary fire during the five days of rioting in Detroit in 1967. Litter and a confused tangle of hose line the street. Joseph Mancinelli, Detroit Fire Department.

she indicated, and then the burning of the city. The entire white population would be murdered, and upon the completion of the task, Caesar, a Negro, would be governor. Hughson and his wife would play important roles in the Macbethlike drama, and Hughson, Mary said, eventually would be crowned king. Many people were arrested purely on Mary Burton's testimony.

According to Mary, a girl named Peggy, "the Newfoundland beauty," was a part of the plot. Peggy, in order to save her life, implicated many other people who were thrown into jail. In the days that followed, Mary, Peggy and Arthur Price, another informer, were always able to come up with some new victim and "their extravagant tales grew with the public terror and excited fresh alarm." Mary's ability in story-telling far exceeded that of the others, and she managed to come up with the names of the three principal Negro "conspirators:" Caesar, Prince and Duffee.

Executions began on May 11 with these three "conspirators." Later, Peggy, Hughson and his wife were also executed. On June 8, six Negroes were chained to a stake and burned. Four more Negroes were sentenced to the same fate on June 10, but one of them "confessed" and implicated still more.

During the entire trial, the prison was so crowded that there was danger of disease. Over 150 people, mostly black slaves, had been imprisoned, and of these, thirteen were burned at the stake, eighteen were hanged, seventy-one were deported, and the rest pardoned or discharged. Blacks appeared before the court without defenders — the best counsels in the city were prosecuting. The guilty verdict was pronounced on victims of "public insanity." New Yorkers, including the most reputable citizens, had plunged into deeds of "unexampled cruelty" in their quest for conspirators.

Cruelty to blacks and black defenders and the use of fire to express that cruelty dot the pages of U.S. history. Mobs roamed the streets of New York City in July of 1863 in a series of draft riots. Rioters "simply desired to break up the draft in some of the upper districts of the city, and destroy the registers in which certain names were enrolled," explained one report.

Fires were set in many buildings, and while the fires were in progress, an attack was made on the Colored Orphan Asylum on Fifth Avenue. As one reporter explained, "There would have been no draft but for the war — there would have been no war but for slavery." Since slaves were black, twisted logic allowed the ravaging of the orphan home. The staff anticipated what was about to happen and directed the children through a rear door just as the mob entered through the front. The furniture and other belongings were smashed and then the frenzied mob set fire to the building. Chief Engineer Decker arrived and pleaded with the mob to let the fire department at the fire. He was knocked down twice in the argument but finally he and two of his men got through to the burning building. Their efforts were valiant but the building was lost.

Blacks were pursued by mobs throughout the city and were tortured, mutilated and lynched. At least

one Negro was roasted to death. A fireman rescued a seven-year-old boy from a mob, but he died from the severe beating a few days later.

Just a little over one hundred years later, in Los Angeles, "the City of Angels," burning was to become the symbol of militant blacks. In the spacious, open California city on the evening of August 11, 1965, people in neat, white neighborhoods watered their lawns. Even in Watts, a predominantly black neighborhood, dwellings were primarily one-family and most had adequate yards, sprinkled that evening with garden hoses. Many people clustered on porches; the temperature was in the eighties and the air was heavy and humid.

A California highway patrolman, making a routine drunk-driving arrest in the neighborhood, attracted a large crowd. The driver was black and a belief that he was treated unfairly was uttered by the crowd, also composed of blacks. The situation was soon out of hand as the crowd formed mobs that stoned and burned cars driven by whites, and then beat the occupants. Molotov cocktails (gasoline-filled bottles with wicks) were thrown at cars and buildings which immediately burst into flame. On the evening of

August 12 violence again erupted, this time with crowds of up to five thousand participating.

By the morning of August 14 rioting was spreading and two thousand National Guard troops began arriving in jeeps with mounted machine guns. One Guardsman said: "They've [the rioters] got weapons and ammo. It's going to be like Vietnam." As fires flared throughout a riot-torn area and as fire equipment dashed to the scene in the traditional way for the purpose of extinguishing the fires, firemen simply doing their duty found that they were not welcome. The fire fighters, once thought of in heroic terms as friends of the community, were fired upon by snipers. A fire set by the rioters, sweeping a three-block section of the city, had to be abandoned as firemen were met with rocks and fire bombs.

Black youths took control of another two-block section and set at least fifteen fires to homes and stores. Observers were quick to point out that most of the damage occurred in the black areas. In the first three days of the riot, 150 cars had been burned, and their smoking frames cluttered the intersections. Over one hundred fires had been set, and an entire 150-block section of Los Angeles was in ruins, with broken store windows and scorched interiors. A thick, choking smoke was suspended over the city.

Arson techniques were improved as the riots continued, and fires were constantly springing up in the riot areas. One black reporter observed a man setting fire to a building after his two companions looted the store. Six gasoline-filled soda bottles were placed inside the broken storefront window and ignited. After each bottle exploded, a popping sound was heard and then, with a sudden roar, the entire building burst into flame. Firemen had been granted permission not to enter areas under gunfire. But the firemen, unable merely to step aside and let Los Angeles burn while they helplessly looked on, fought their way to the scene of the fire; it took them an hour to get there. As they climbed ladders into the fire, rioters threw bricks at them and taunted, "Burn, baby, burn!"

Fire department officials estimated damage in the millions of dollars. The fire situation was described as critical — neighborhoods were completely engulfed in flames. Some fire companies succeeded in reaching only a few of the fires, and many fires had to be allowed to rage out of control. One big fire completely devastated a furniture factory and spread from there to five nearby black homes. Firemen arriving on the scene were greeted with Molotov cocktails and rocks.

Patterns began to appear. Invariably, stores were looted, then burned. After setting the fires, rioters waited for the arrival of the fire equipment and their former friends, the firemen, and then hurled missiles at them. Often neighborhood residents devised their own means of fire fighting. One black man used the puny stream of his garden hose in at attempt to quench a raging fire that enveloped four stores in the center of the riot area.

President Lyndon Johnson called the riots "tragic and shocking." Los Angeles Mayor Samuel Yorty used the adjectives "terrifying" and "hysterical" to describe the riot-swept sections of the city. The death toll continued to climb. One of those deaths was a

(continued on page 182)

Overleaf: *Flames glow in the riot-heavy night air on Chicago's West Side in April 1968.* Chicago Daily News *photo by Jim Klepitsch.*

firemen, killed by a falling wall at the scene of an incendiary fire.

During the day of August 14, a fire captain inspected the area. At least twenty major fires burned throughout the day, and over two hundred had been reported. Fire fighters endangered themselves whenever they went in to fight them and, consequently, were often forced to leave. Even if they did not have to endanger themselves by being under attack, there was a lack of personnel to handle all the fires. "We simply don't have the manpower to extinguish them," explained Fire Captain William Clutterham. He toured the area in a police helicopter and said, "From the air, I could see them [the rioters] lighting new fires by throwing inflammable liquids against the sides of buildings." The captain had been at the scene of many of the major fires and reported that whole groups of buildings were often involved. "Many times by the time we get there they have pretty good starts," he explained. "Our purpose is merely to keep them from spreading to private houses."

The skyline was obscured by smoke and the streets were almost entirely deserted. Nevertheless, police set up roadblocks to protect the firemen. Equipment raced throughout the city as the firemen continually attempted to battle blazes, often in the same places they had fought them before. Water from their equipment constantly flowed down sidewalks and streets, trickling into gutters. On one street alone — 103rd Street in Watts — fourteen separate fires raged at one time. The area of East 103rd Street was finally left to burn when a fireman was wounded by a sniper's bullet. One person on the fire department described three fire houses as being "under siege." As fire crews left the station house they were often subject to gunfire. At least thirty-two of those injured in the riot were firemen. Later, two hundred "flak suits" made of bulletproof mesh armor were procured for the firemen since it was realized that only the fire fighters stood between saving Los Angeles and letting it become a city in ashes.

By the evening of August 14, a curfew from 8:00 p.m. to sunrise was in effect. The riot had at first involved only a few square blocks of the low-income Watts area but quickly spread until by curfew that evening it covered 21 square miles out of a total of Los Angeles' 457 square miles. But it was not yet over. The black proprietor of a looted shoe store said,

"I'm afraid it's going to spread before it finally quiets down."

The injured civilians within Watts were often unable to flee the area. Ambulances tried to get through but they, too, were met with a shower of bricks and Molotov cocktails. Many in the riot-torn areas were hungry since supply trucks were unable to get through. Governor Edmund Brown ordered the disaster office to distribute food. The use of fire as a revenge weapon was frighteningly effective and the governor, upon touring the ruins as violence began to subside, remarked, "The trouble is that five minutes after we leave somebody could come in here and start a fire."

By August 15, four days after the riots began, people were weary and the National Guard was irritable and exasperated. A few fires still burned and Central Avenue was blocked by fire hose. A police helicopter directed the firemen in the fire-fighting efforts which still took place amid gunfire. A combatant spirit arose and a sign erected by the National Guard at Wilmington Avenue and 104th Street read "Turn Left or Get Shot." The Guard had been assigned to ride on top of the city's fire equipment as it went into riot areas. By the time the curfew was lifted on August 17, thirty-four people were dead, 1,032 were injured, four thousand were arrested and property damage by fires and looting amounted to thirty-five million dollars.

The riot that had started from a small incident on a warm summer night and evolved into a disruption that adopted "Burn, baby, burn" as its unofficial theme had found support among all economic classes of blacks, whether they actually participated or not. One Los Angeles black businesswoman explained, "I will not take a Molotov cocktail but I am as mad as they are." A similar pattern emerged nearly two years later in Newark, New Jersey.

Newark was founded in 1666 by a group of eighty-four Puritans who left Connecticut in order to practice their religious beliefs. In 1967, Newark's population was 400,000, about half of which was black. Racial tension had been increasing — the black population wanted their own candidate, a man they claimed had experience, to be secretary of the school board, rather than the mayor's choice. Also disputed was a fifty-acre tract of land in the predominantly black Central Ward that had been chosen for a

Medical College. Black leaders contended that it should be used for housing.

On July 12, 1967, a black cab driver was arrested on charges of assaulting a policeman. A crowd of two hundred blacks chanted "Police brutality!" in front of the Fourth Precinct station house, where the cab driver was held. Shortly after midnight the fire department received several false alarms, and their equipment could be heard roaring throughout the area. In the early morning hours of July 13, stores were firebombed and looted. At 2:20 a.m. on July 14, Newark Mayor Hugh J. Addonizio telephoned Governor Richard J. Hughes and requested troops. The mayor described the Newark situation as "ominous."

A large fire, which the fire department said had not been set by the rioters, raged at the intersection of Broad and Market streets in the center of downtown Newark at the beginning of the disturbances. Police accompanied the fire equipment to the scene, toting shotguns to protect the firemen. A luggage store and jewelry shop were gutted and the blaze drew thousands of black spectators. Some spectators broke away from the crowd as the fire continued its path through the business section, and they smashed windows and looted. Police shot their guns through the air to disperse the crowds. Then a store on Broad Street was firebombed and a toy store was burned by an exploding Molotov cocktail. The blazes spread to engulf other stores.

A white fire captain, thirty-eight-year-old Michael Moran, was killed by a sniper while fighting a blaze in the riot area on July 15. The Uniformed Fire Officers Association of New York City offered a one-thousand-dollar reward leading to the arrest of the murderer. At least thirty firemen were injured by gunfire throughout the rioting. Governor Hughes was appalled at the violence and commented: "The people of Newark have to choose sides. They are either citizens of America or criminals who would shoot down a fire captain in the back. . . ." Firemen were often under heavy sniper fire; one commented, "The fires weren't the problem — it was the jeering and snipers' shots."

Later that same month Detroit experienced the worst riot in the nation's history. It lasted for five days and killed forty-one people. On July 23, police raided an illegal, all-night tavern on Twelfth Street, in

(continued on page 186)

Fire rages through luggage and jewelry stores in downtown Newark (below) at the beginning of the 1967 riot. Wide World. A Detroit fire chief sighs with exasperation as police protect the working fire fighters (opposite, top). Detroit Fire Department. The firemen had many close calls as snipers shot at them while they were responding to the alarms (opposite, bottom). James Haight.

A National Guardsman with ready bayonet maintains a firm stance (above) as Chicago firemen perform their tasks in the background. Wide World. Causes of a riot may still smolder but the riot itself ends (opposite) and people awake to loss of life and property. Joseph Mancinelli, Detroit Fire Department.

the heart of the black neighborhood. During the raid there were accusations and arguments, and someone placed a quick call for more paddy wagons. The riot was on.

Shortly after it began, a black college student rushed to the Twelfth Street area, hoping to witness a true revolt of his people. Upon his arrival, however, he was disappointed to see people involved in looting rather than ideology. A fire chief at the scene implored for help from the crowd and the black college student volunteered for duty. He was the only black that worked with the fire department crew during four days of combating arsonists' blazes.

Thousands participated in the firebombing and looting. The National Guard was called out by Governor George Romney as the warm wind fanned the many fires. A blaze in one area spread a solid sheet of flames over a ten-block section of the city. Tenements and businesses burned in the near West Side as more fires broke out on the Northwest Side. Looting and firebombing continued throughout the Twelfth Street area. One group of looters ran past the tired firemen and dropped off two six-packs of beer in a sarcastic gesture. Firemen at one point were forced to leave their duty since they did not have any protection from the bricks, bottles and other objects

hurled at them. They left their hose in the streets. However, when they later returned, the residents of the burning areas offered protection. About twenty blacks on one block took up rifles and surrounded the firemen to protect them from their attackers. "They say they need more protection," said one black, "and we're damn well going to give it to them." In the first two days of rioting, 731 fires had been reported and smoke stifled the city.

Black leaders condemned the officials for not gaining control and taking action sooner. Many blacks were forced to leave their homes and seek shelter with friends or the Salvation Army. By July 28, when the violence was nearly over, 1,163 fires had been reported and a few still burned. The people that had been burned out of their homes were given lists of available housing, food distribution depots and grocery stores.

Dr. Martin Luther King, Jr., to many blacks a symbol of hope that "we shall overcome," was killed in Memphis on April 4, 1968, by an assassin's bullet. Riots broke out in many cities following the tragedy. Washington, D.C., and Chicago were among the hardest hit.

Immediately after Dr. King's death was made known, protest marches were organized in Washing-

ton which escalated to full-scale riots. President Johnson called out 6,700 U.S. Army and National Guard troops on April 5; some of the troops guarded the Capitol and the White House. Black Mayor Walter E. Washington declared a curfew from 5:30 p.m. to 6:30 a.m. The police force of 2,800 lacked the manpower to respond to looting and burglary calls, plus protect motorists and firemen. Government workers had been advised to leave the city and by midafternoon Washington was engulfed in a massive traffic jam. Fire apparatus, answering an unending rush of fire calls, was caught in the tie-up and could not get to the fires.

Most of the burning and looting throughout the disturbance was confined to black areas, although two large department stores in the downtown area were pillaged and set on fire. Suburban volunteer fire departments donated fifty pieces of fire equipment to aid the overburdened Washington department. The fires came within two blocks of the White House before the rioting began to decrease. By Sunday, April 7, all was quiet. At least seventeen firemen had been injured and more than seven hundred fires had been started.

In Chicago, all off-duty firemen were ordered to their stations. When the trouble began just after Dr. King's death, firemen used more than one hundred pieces of equipment to fight blazes on the black West Side. Fires cut off electricity and there was no phone service, as telephone poles burned to the ground. Several stores ignited at once while residents chanted to police and firemen, "A white man killed Martin Luther King." Fire Commissioner Robert J. Quinn directed the battle against the fires by helicopter.

As the hours went on, fires continued to be reported. Firemen faced the peril of snipers' bullets while fighting a fire on April 6 on the near North Side. By the end of the day, Chicago had suffered 125 major fires, and many of the buildings had been burned to the ground. The West Side was covered with smoke. Firemen were constantly harassed in their attempt to put out the fires: Angry rioters continually opened the fire hydrants in the disrupted areas, and when fire fighters arrived to close the hydrants, they were attacked by thrown objects or snipers' bullets. But the firemen succeeded at extinguishing most of the blazes and then watered down the twenty-eight block area. By April 7 there were only sporadic incidents of violence and many of the firemen, some of whom had worked thirty-six hours without sleep, began to leave. Over one thousand people had been burned out of their homes.

The riots that occurred after Dr. King's death differed from the Los Angeles, Newark and Detroit riots in that they were spontaneous eruptions that were motivated by a common grief rather than triggered by a particular event occurring in a certain section of the city. Areas hit by riots did not quickly recover — many blacks felt that their condition of living was such that they had not lost anything, even if their homes had burned to the ground. The characteristic optimism of rebuilding exhibited after other major American fires was not present. One black said of his house: "Mine ain't worth a damn nohow." And American Negroes sought to "emancipate themselves by violence."

Firemen to the Rescue

*Firemen have a simple motto: "Fire fighters fight fires and save lives." While
it is important to prevent property damage, a human life is irreplaceable. In the fireman's
search for trapped victims he sometimes finds them too late. Though it is all in a day's
work, rescue somehow never becomes routine. Firemen often find themselves in precarious
positions at the scene of a fire (below) and must call upon their fellow firemen for
help. Chicago Fire Department. A fire lieutenant's expression (opposite) indicates his
involvement with saving lives as he removes a young victim. Wide World.*

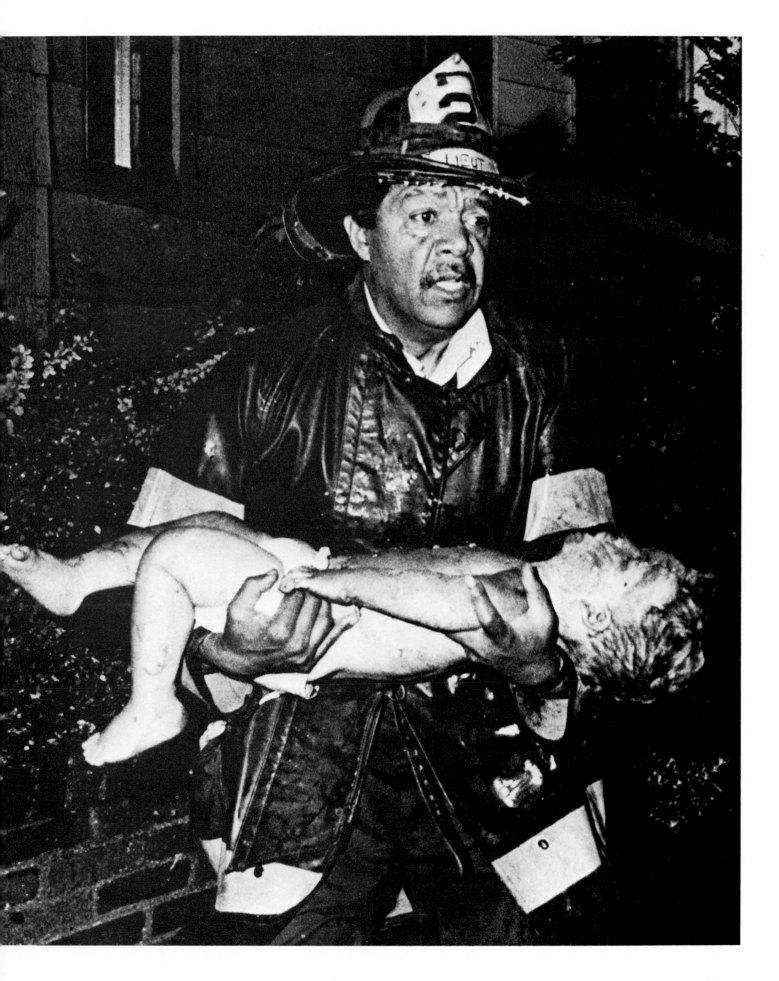

Children are some of the most helpless victims of fires and the most tragic because they have so much to live for. A fireman tries to breathe life into a young boy (above). Wide World. With grave concern etched on his face, a fire fighter rushes a pajama-clad child to an ambulance (opposite). Milwaukee Journal from James Haight. Other children were lucky: They were rescued by firemen (right) before they were overcome by choking fumes. Chicago Fire Department.

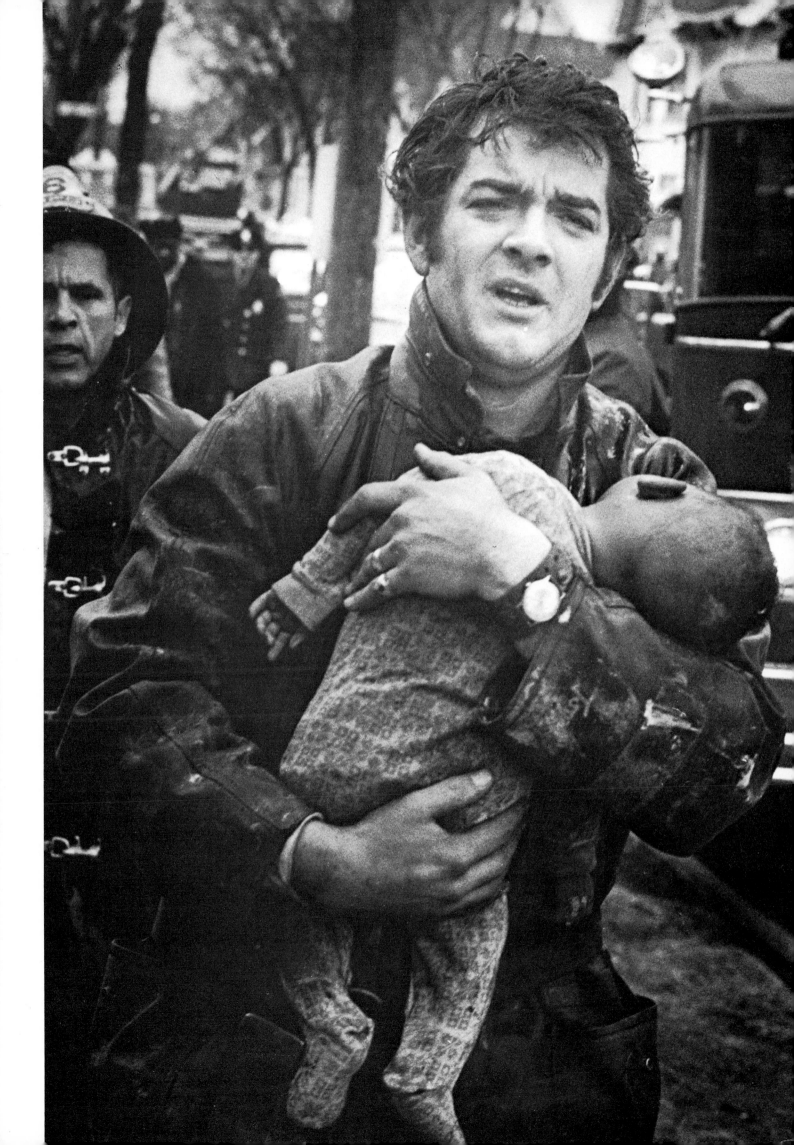

Firemen in a Modern America

"Anyone smell smoke?" asked a fireman at the scene of what turned out to be a false alarm. A mutual shrug from his companions and all promptly returned to the station house for a meal of hot dogs and baked beans, a meal they had chipped in for. Some, while on the way back aboard the shiny red truck, surveyed the precious, efficient equipment, much of which had been purchased with money from their own pockets.

"They're saying it's a fireman's fault that there aren't more fires," said a fire fighter who felt his services and frustrations went unappreciated. And, while less than one percent of the fires today spread beyond the immediate area, a fireman's job involves all the same dangers and risks that it always has. "Look," the fire fighter continued, "we're talking about saving lives and a life can't be replaced."

"The biggest frustration is on the street," said another fireman. "You go into a place for a rescue and get half-killed. It hurts, especially when you're giving it all you got." Fire fighters have become "authority symbols" of a status quo society, and their occupation, the world's most dangerous, and their involvement with life and its preservation have been set aside and forgotten. One fire fighter recalled saving the life of a child in a burning building only to return to the same neighborhood at another time to be greeted with rock-throwing and name-calling, some of which was done by the very child he had saved.

"The city doesn't know what citizens it has in the firemen," said another. "Where else are you going to find guys so dedicated they're happy to work harder?" While the forty-hour week has become as American as apple pie, many fire departments require an average week of fifty-five hours. New York State has legislated a forty-hour work week by 1974 for its fire fighters.

Flames lick at the framework of a home and throw tongues of fire into the evening sky (opposite). Robert Bartosz. A fireman prepares to descend a ladder as smoke pushes from behind (right). Milwaukee Sentinel.

A small flame can turn into a bright, roaring inferno in a matter of minutes, destroying life and property. James Haight.

Smoke curls from windows as a fire fighter struggles to lift hose to an upper story of a burning building while his partner waits. Robert Bartosz.

The dedicated, well-trained fire horses were retired to farms years ago, though some horse-drawn apparatus could be seen in some cities in the 1920's. Fire fighting had become their business, too, and their ears still twitched at the sound of an alarm as they listened from the serenity of their pastures. The firemen who knew them and grew to love them always maintained that the horses wished they still were going to every alarm. The spirited horses had been replaced by a new device called the internal combustion engine. By 1915 the clomping of horses' feet was being supplanted by the inanimate, gasoline-powered tractors.

Fire dogs, usually traditional dalmatians, responded well to the calling of the fire service and dutifully ran ahead of the horses and artistically gilded steamers, clearing the way. The dogs, however, could not outrun the efficient motorized vehicles with their blaring sirens and standardized red paint that were unquestioningly given the right of way on even the busiest of streets. Fire dogs found other jobs around the station house such as riding the equipment to a fire and then guarding it, or staying at the

station house to prevent burglary when the firemen were on a run.

The fire service was becoming more scientific in its approach to fires and modern innovative equipment required serious, intelligent study and knowledge. This fast-changing technology made some vocational colleges add courses leading to an Associate Degree in Applied Science of Fire Technology. Firemen today are highly skilled and must pass vigorous tests before they are installed into a fire department. A fireman is a chemist, engineer, mechanic and electrician, in addition to "simply" fighting fires.

But scientific equipment has not changed the composition, degree and intensity of fires. Though the methods of dealing with them have changed, fires themselves have not, and it is still the same risky business to enter a burning building. A dwelling fire today is just as intense and causes the same anguish that it did in Boston in the seventeenth century. A collapsing roof or a falling beam kills today as it did then. Fire fighters emerge from a smoke-filled building today, coughing and choking as they did in early Jamestown. Still, when occupants of a building being

A fire fighter, dripping wet, gasps for air (right) at a window. Repeated exposure to the products of combustion shortens a fireman's life by an average of more than eight and a half years. Robert Bartosz. Though a rainbow may appear in the mist (opposite), a fireman's real reward comes with knowing the fire's out. James Haight.

ravaged by fire are running out, the fire fighter is running in. His reasons are many: He cannot stand the wanton, indiscriminate destruction of life by fire, or the waste of hard-earned property through fire, whatever the cause. He battles fires whether caused by the vengeance of a pyromaniac or by an over-heated boiler or a spark from a home-owner's burning pile of fallen leaves or weeds. Acts of heroism are everyday events in the life of a fire fighter:

— A fireman battling a blaze in a furniture store is dazed by an explosion but still fights to save the life of an injured fellow fireman.

— A house is on fire on a winter's day in early 1970 and a family is trapped inside. Fire fighters grope through the extreme heat and heavy smoke to perform rescue operations. Long after others would have quit, these men find endurance to continue; their reward is the lives they save.

— A fire fighter's response to life and its preservation becomes instinctive. An off-duty fireman monitors a fire call radio. Upon discovering the catastrophe is near his home, he quickly responds and gives first aid to a neighbor's child before arrival of the rescue squad.

— A woman on the fifth floor of a building threatens suicide. The fire department is called and it is the imagination and efforts of a fire fighter that bring her down the ladder to safety.

— A window washer perilously hangs from an upper-story window of a tall building. The instinctive action of a fireman at the scene is to grasp the victim and help him to safety.

— Another building is on fire and a team of fire fighters removes an occupant through the skillful use of ladder equipment. Another person is still inside and they must enter the building, which is burning with great intensity. The fire fighters don their masks and proceed into the high heat and choking smoke, but they are successful in removing the second-floor occupant.

— A vacationing fireman is the first on the scene at a restaurant fire. By the time the local fire department arrives, he has the blaze under control.

— An off-duty fireman gives mouth-to-mouth resuscitation to a child that would have otherwise died before the arrival of the rescue squad.

— Another off-duty fireman witnesses an automobile collision and rushes to give the victims first aid and also successfully prevents a large fire from a ruptured gasoline tank.

— The fire department is called to a burning house in 1972 and in a little girl's own words, "Smoke was coming toward my face. . . . The man with the funny long nose and glass over his eyes . . . put the ladder up to the window and took me down."

— Firemen dedicated to saving lives attend an eighty-hour emergency medical technician course at a hospital, without pay. When a fire later breaks out in

a flat, fire fighters search the flaming, smoking structure for twelve minutes in an effort to find a two-year-old baby. They find the baby, but it appears to be lifeless. Two firemen, graduates of the course, administer heart massage and mouth-to-mouth resuscitation. Life returns to the child.

— Vandals set fire to rubbish on the first floor of a vacant building, and it quickly spreads to the third floor. When the fire department arrives, two firemen are injured when they fall while hauling a hose line from the second to the third floor. Hose when filled with water is heavy, weighing about 150 pounds per fifty-foot length.

—While fighting a blaze at a popular restaurant in the Chicago Loop in early 1973, three firemen are killed. "The men were in there working when, without warning, the roof caved," Fire Commissioner Robert Quinn said.

From the first American fires that were fought by bucket brigades after a fire's discovery, following cries of "Throw out your buckets!" to the modern

(continued on page 202)

Overleaf: *Yellow flames merge into great orange clouds, silhouetting the firemen who are hauling hose to the blaze. Robert Bartosz.*

197

Outlines of fire fighters doing their job are carved from a sheet of flame that crackles through a wooden structure. Even the early evening sky is blotted out by flames. James Haight.

Chemical foam is used to blanket certain types of fires and occasionally also blankets the fireman (left). A fire fighter (below) takes time out to acknowledge a long-time companion in combating fires, the dalmatian. Both James Haight.

citizen who, upon sniffing the air and smelling smoke, exclaims, "Call the fire department!" fires and fire fighting have been an exciting but wastefully destructive everyday feature of American living.

In the first American settlements, all "able-bodied" men were required to turn out for a fire. They turned out for the numerous Boston fires of the seventeenth and eighteenth centuries. While it was recognized that a small blaze could soon get out of hand, the size of the blaze, it was soon discovered, had nothing to do with its cruel ability to take a life. Many early fire fighters were killed in their attempts to protect the property and lives of others.

Firemen, tired and overworked, responded on a winter night in New York City in 1835 with a temperature of zero degrees and worked with valiant perseverence until they stopped the fire. Even with all the odds distinctly against them, they fought with "superhuman" strength in a large fire that destroyed most of Portland, Maine, in 1866.

Tired Chicago firemen, many awakened from a restless sleep, somehow anticipating an impending tragedy, responded with a shortage of equipment but a large amount of courage to a fire in the O'Leary barn in 1871. The same evening, more than two hundred miles to the north at Peshtigo, volunteer fire

people were killed, including the entire volunteer fire department. Firemen, once considered by everyone to be friends of the people, were fired upon in Los Angeles, Newark, Detroit, Chicago and other cities in the tumultuous 1960's, and still they insisted on risking their lives to perform their duties.

An average of 120 fire fighters are killed each year. A fireman's life span is shortened by breathing carbon monoxide and inhaling smoke and toxic gases. He fights hostile fires throughout the country every day and is willing to give the supreme sacrifice — his life — for the benefit of others.

While polishing his sleek, modern machine, he may tell the little boy who wandered into the station that fire prevention in the future, by the time the boy is ready to begin fire fighter training, will be even more technologically advanced and automated than now, just as a fireman had told him when he was growing up. Better equipment and more technological knowledge has been a traditional objective of fire departments since the first organized company in 1736.

The struggle between humanity and fire still goes on. The romance of the fire department was not merely in the red shirts, the decorated fire apparatus, the dashing horses. It rests with the men and their job which is the same today as it was over two hundred years ago.

Fire fighters are subjected to frequent injury and sometimes hostile acts from the very people they protect, but they do not hesitate in the performance of their duties. They work hard and unendingly to save life and property, and to put out the fire, and do it with courage, without any expectation of recognition. Their eyes water from the smoke, their lungs fill with the poison of products of combustion, and they search for trapped life at the risk of their own. Yet they continue to enter into burning buildings amidst the brilliance and heat of flames day after day, because it's their job.

fighters moved their proud little fire engine, the "Black Hawk," into position to fight the forest fire that would go down in history as America's greatest natural disaster.

Fire Chief Edward Croker of New York had repeatedly warned against such factory firetraps as the Triangle Shirtwaist Factory, in which 145 people were killed in 1911, despite heroic efforts of the fire department.

Fire fighters were on the scene in 1942 at Boston's Cocoanut Grove fire, breaking into the still-burning structure and searching for signs of life. In the Texas City explosion in 1949 an estimated one thousand

Sober-faced firemen at the gravesite (above) honor a colleague killed
in the line of duty. Joseph Mancinelli, Detroit Fire Department. Other fire
fighters (below) salute their fallen comrade. James Haight. An average of
120 firemen die each year of fire-related injuries. But despite the danger,
firemen continue to battle fires (opposite). James Haight.

Fire Museums of the U.S.

The following list is not intended to be complete, but only a brief guide to museums
containing fire memorabilia throughout the United States.

Alabama

The Phoenix Fire Museum
203 South Claiborne Street
Mobile, Albama 36602

The museum is located in the restored fire house of the
Phoenix No. 6 Volunteer Company and contains steamers
from the 1890's, a motorized ladder truck, a helmet collection
and a large trumpet collection.

Arizona

The Hall of Flame
3626 Civic Center Plaza
Phoenix, Arizona 85251

The museum's artifacts include a huge collection of antique
fire-fighting equipment from the nineteenth and twentieth
centuries, fire trumpets, fire helmets, fire marks and fire
memorabilia from all over the world.

*The last hand pumpers were built around 1859, just
prior to the heyday of steam engines. James Haight.*

California

Old Plaza Fire House
134 Plaza Street
Los Angeles, California 90012

Located in the building that was home to Los Angeles's first
fire company, the museum contains hand-drawn and horse-
drawn equipment, a fire alarm display, photographs, helmets
and badges.

Pioneer Hook and Ladder Company
1572 Columbia Street
San Diego, California 92101

Housed in an old fire station, the museum contains equipment
from 1841 to 1940, badges and photographs.

San Francisco Fire Department Pioneer Memorial Museum
655 Presidio Avenue
San Francisco, California 94115

The museum has hand-drawn, horse-drawn and motorized
equipment and a collection of uniforms and photographs.

Connecticut

Connecticut Antique Fire Apparatus Association
Scantic Road
Warehouse Point, Connecticut 06088

Among this museum's collection is a group of motorized fire
trucks dating from 1926 to 1949.

District Of Columbia

Fire Department Museum
4930 Connecticut Avenue N. W.
Washington, D. C. 20008

The Friendship Fire Association maintains the museum which
contains an old steam engine, buckles and firemen's costumes.

National Museum of History and Technology
Division of Transportation
Smithsonian Institution
Constitution Avenue at 14th Street N. W.
Washington, D. C. 20560

Part of this huge collection is devoted to fire-fighting equipment and includes a hand pumper built around 1740 and
a motorized engine from 1920.

Florida

Jacksonville Fire Department Museum
12 Catharine Street
Jacksonville, Florida 32205

This museum is scheduled to open in the near future and will
be located in the same building that belonged to No. 3 Station,
which closed in 1928. It will include a motorized 1912 hook-
and-ladder truck, fire marks, helmets, trumpets and buckets.

Illinois

Chicago Historical Society
North Avenue and Clark Street
Chicago, Illinois 60614

The Fire Gallery section of the museum contains numerous
artifacts that pertain to the Great Chicago Fire of 1871.
Included are photographs, a slide presentation, hand-drawn
apparatus, models of horse-drawn apparatus, a working diorama
of the Chicago fire and various Chicago fire relics.

Louisiana

Presbytere Museum
751 Chartres Street
New Orleans, Louisiana 70116

The museum has a fire exhibit which contains items pertaining
to the fire service between the years 1820 to 1900. Both
hand-drawn and horse-drawn apparatus are included plus helmets, fire marks, photographs, banners, fire tools and others.

Maine

Lincoln County Fire Museum
Federal Street
Wiscasset, Maine 04578

The collection contains fire equipment dating from 1803.

Maryland

Fire Museum of Maryland
1301 York Road
Lutherville, Maryland 21093

The museum contains forty pieces of fire equipment dating to the beginning of the nineteenth century. Also on display is a telegraph alarm system, a helmet collection and other fire memorabilia.

Massachusetts

New England Fire And History Museum
Route 6a
Brewster, Massachusetts 02631

In addition to a working diorama of the Great Chicago Fire, the museum has a collection of antique fire-fighting equipment, axes, leather buckets, old fire hose, prints and fire marks.

National Fire Museum, Inc.
21 Endicott Street
Newton Highlands, Massachusetts 02161

A collection of antique fire equipment, fire helmets and fire engine models is included in the museum.

Michigan

Henry Ford Museum
Greenfield Village
20900 Oakwood Boulevard
Dearborn, Michigan 48121

An old fire house located on the grounds contains equipment dating to the 1790's. In addition to the hand-drawn, horse-drawn and motorized apparatus, the museum displays lanterns, fire axes, badges and convention ribbons.

Missouri

Missouri Historical Society
Jefferson Memorial Building
Lindell and Debaliviere
St. Louis, Missouri 63112

Hand-drawn equipment, helmets, paintings of famous firemen, trumpets, fire hose and photographs are included.

Nebraska

Pioneer Village
Highways 6 and 10
Minden, Nebraska 68959

The collection includes fire-fighting equipment from 1830 to the present in a restored fire house.

New Jersey

Museum of the Newark Fire Department Historical Association
49 Washington Street
Newark, New Jersey 07101

Displays include fire engines, helmets, photographs, paintings and documents.

Before motorized equipment, about 1917, northern firemen relied on skis in winter. James Haight.

Trenton Firemen's Museum
244 Perry Street
Trenton, New Jersey 08618

This museum, housed in the old firemen's gymnasium, contains thousands of items that pertain to the fire service, including photographs, helmets and hand-drawn equipment.

New York

Ye Olde Fire Station Museum
8662 Cicero - Brewerton Road
Cicero, New York 13039

The museum contains hand-drawn, horse-drawn and motorized equipment. It also has a sizable collection of fire helmets, all types of glass hand extinguishers, wall-hung extinguishers and fire badges.

American Museum of Firefighting
Harry Howard Avenue
Hudson, New York 12534

The collection contains thirty-five pieces of old fire equipment dating from 1731 to 1926 and numerous paintings and prints.

Firefighting Museum of the Home Insurance Company
59 Maiden Lane
New York, New York 10038

The company's museum includes a reconstructed fire house with displays of fire fighter costumes and tools. The large and complete collection includes over three thousand fire marks, old fire equipment, hat fronts, badges and photographs.

New York Fire Department Museum
104 Duane Street
New York, New York 10007

The collection includes apparatus from 1820 to early motorized equipment, plus leather buckets, trumpets, bells, medals, photographs, prints, fire engine models and alarm boxes.

Museum of the City of New York
1220 Fifth Avenue
New York, New York 10029

A hand-drawn fire engine, possibly used in the New York fire of 1835, is on display. The museum also includes prints, paintings, fire engine models, hats, trumpets, auxiliary equipment and restored fire engines.

Ohio

Allen County Historical Museum
620 West Market Street
Lima, Ohio 45801

The museum has hand-drawn, horse-drawn and motorized equipment. Also on display are fire extinguishers, helmets, a trumpet, a fire bell, axes and other items.

Oklahoma

Oklahoma State Firefighter's Museum
2716 N. E. 50th Street
Oklahoma City, Oklahoma 73111

Exhibits include eighteen antique fire engines, a fire alarm dispatcher of World War I vintage, a display of early fire extinguishers, and the first fire house in the State of Oklahoma.

Pennsylvania

Hershey Museum
Hershey, Pennsylvania 17033

Hand-drawn and horse-drawn equipment is featured, including an engine that was manufactured in London around 1764.

Fire chiefs had their own buggies to take them to fires, and they usually arrived first. James Haight.

Insurance Company of North America
1600 Arch Street
Philadelphia, Pennsylvania 19101

The company's collection contains fire engines, engine models, prints, paintings and nineteenth-century historical material relating to the fire serivce.

Philadelphia Fire Department Museum
149 North Second Street
Philadelphia, Pennsylvania 19102

Included are hand-drawn and horse-drawn fire engines, helmets from all over the world, hand-carved engine models, trumpets and auxiliary equipment.

Tennessee

Memphis Fire Department Museum
2668 Avery Street
Memphis, Tennessee 38112

The museum has displays of antique equipment, apparatus models, photographs and fire department relics.

Texas

Dallas Firefighters Museum
3801 Parry Street
Dallas, Texas 75226

The museum's collection is housed in the oldest fire station in Dallas and consists of documents, hand tools, hand-drawn hose carts, horse-drawn steamers, motorized engines and fire extinguishers.

Thomas W. Hopkins and Son Insurance Local Agents
2033 Norfolk Street
Houston, Texas 77006

Displayed are: firemen's belts, flags, uniforms, membership certificates, fire marks, buckets, helmets, parade hats and fire engine lanterns.

Virginia

Friendship Veterans Fire Engine Company
107 South Alfred Street
Alexandria, Virginia 22314

This museum contains the engine that was presented by George Washington to the volunteer company in 1775. The collection includes other hand-drawn and horse-drawn equipment from the eighteenth and nineteenth centuries and a display of fire marks.

Washington

Washington State Fire Service Historical Museum, Inc.
Seattle Center
P. O. Box 9521
Seattle, Washington 98109

Hand-drawn, horse-drawn and motorized fire engines are displayed, plus a fire boat display and an operating fire alarm.

Wisconsin

Milwaukee County Historical Society
910 North Third Street
Milwaukee, Wisconsin 53203

Among the items included is a hand-drawn pumper from 1850, a hand-drawn hose cart, artifacts from Milwaukee's fires, including the Newhall fire, fire helmets from 1846 to the present, fire extinguishers and photographs.

Peshtigo Fire Museum
400 Oconto Avenue
Peshtigo, Wisconsin 54157

Included in this general history museum are items that managed to survive the disastrous Peshtigo fire.